Atomic, Molecular, and Optical Physics Handbook

Atomic, Molecular, and Optical Physics Handbook

Contributors

Martin Veis and Roman Antos et al.

AURIS
Reference

www.aurisreference.com

Atomic, Molecular, and Optical Physics Handbook

Contributors: Martin Veis and Roman Antos et al.

Published by Auris Reference Limited

www.aurisreference.com

United Kingdom

Atomic, Molecular, and Optical Physics Handbook

ISBN: 978-1-78154-890-5

British Library Cataloguing in Publication Data
A CIP record for this book is available from the British Library

Printed in the United Kingdom
Exclusively distributed by CBS Publishers & Distributors Pvt. Ltd.
Sales & Distribution Rights only for India, Pakistan, Bangladesh, Sri Lanka, Nepal and Bhutan. This book is not to be sold outside these territories.

Contents

List of Abbreviations

AO	Adaptive optics
AO-SLO	Adaptive optics scanning laser ophthalmoscopy
AFM	Atomic force microscopy
BP	Bandpass filter
NIST	CODATA
CM	cold mirror
CD	Critical dimensions
AO	daptive optics
DM	Deformable mirror
DLP	Digital Light Processing
EDAX	Energy-dispersive X-ray diffraction
EDFA	Especially, erbium-doped fiber amplifiers
FOV	Field of view
FFT	Fourier transform
FAZ	Foveal avascular zone
GHS	Gas handling system
GS	Guide star
HH	High-order harmonics
LA	Lenslet array
LWR	Line width roughness
LCA	Longitudinal chromatic aberration
LLR	Lunar Laser Ranging
MCP	Microchannel-plate
MBMR	Molecular beam magnetic resonant
NGS	Naturally existed object
NLO	Nonlinear optical
Oct	Optical Coherence Tomography
PCDM	PhotoControlled Deformable Mirror
PoC	Proof-of-Concept
RQP	Relativized Quantum approach
RNFL	Retinal nerve fiber layer
SNOM	Scanning near field optical microscopy
SP/MAS	single pulse magic angle spinning
SCF	Supercritical fluid
SLD	Super-luminescent diode
SRC	Synchrotron Radiation Center
QP	Quantum Physics
WFA	Wavefront analyzer

List of Contributors

Martin Veis
Institute of Physics, Faculty of Mathematics and Physics, Charles University
Institute of Biophysics and Informatics, 1st Faculty of Medicine, Charles University

Roman Antos
Institute of Physics, Faculty of Mathematics and Physics, Charles University
Czech Republic

Hong Duc Doan
Department of Mechanical and Control Engineering, Tokyo Institute of Technology, Meguro-ku, Tokyo,, Japan

Kazuyoshi Fushinobu
Department of Mechanical and Control Engineering, Tokyo Institute of Technology, Meguro-ku, Tokyo,, Japan

Kyung M. Choi
University of California at Irvine, USA

Luxi Li, Xianbo Shi
Brookhaven National Laboratory, Upton, NY, USA

Cherice M. Evans
Department of Chemistry, Queens College – CUNY and the Graduate Center – CUNY, New York, NY, USA

Gary L. Findley
Chemistry Department, University of Louisiana at Monroe, Monroe, LA, USA

Walter J. Christensen Jr.
Department of Physics and Astronomy, Cal Poly Pomona University, Pomona, USA
Department of Physics, Cal State Fullerton, Fullerton, USA

Zilong Kong
Wufangfa Computer Technology Service Department, Guangzhou, China

Manuel Dorado
CIRTA, Rotation and Torque Research Center, Theoretical Group, Madrid, Spain

Zoran Popovic
Department of Ophthalmology, University of Gothenburg, Gothenburg, Sweden

Jörgen Thaung
Department of Ophthalmology, University of Gothenburg, Gothenburg, Sweden

Per Knutsson
Department of Ophthalmology, University of Gothenburg, Gothenburg, Sweden

Mette Owner-Petersen
Retired from the Telescope Group, Lund University, Lund, Sweden

S. Bonora
CNR-IFN, Laboratory for UV and X-Ray and Optical Research, Padova, Italy

R.J. Zawadzki
VSRI, Department of Ophthalmology and Vision Science, University of California Davis, Sacramento, CA, USA

G. Naletto
CNR-IFN, Laboratory for UV and X-Ray and Optical Research, Padova, Italy
Department of Information Engineering, University of Padova, Padova, Italy

U. Bortolozzo
INLN, Université de Nice-Sophia Antipolis, CNRS, France

S. Residori
INLN, Université de Nice-Sophia Antipolis, CNRS, France

Jingyuan Chen
Yunnan Astronomical Observatory, Chinese Academy of Science, China

Xiang Chang
Yunnan Astronomical Observatory, Chinese Academy of Science, China

Jean-Paul Auffray
Ex Courant Institute of Mathematical Sciences, New York University, New York, NY, USA

Mohamed S. El Naschie
Department of Physics, University of Alexandria, Alexandria, Egypt

Preface

The text *Atomic, Molecular, and Optical Physics Handbook* comprises a comprehensive reference source that unifies the entire fields of atomic, molecular, and optical physics, assembling the principal ideas, techniques and results of the field. First chapter focuses on atomic force microscopy in optical imaging. Second chapter deals with fluidic optical devices based on thermal lens effect. In third chapter, the nonlinear optical (NLO) properties of the doped xerogel film have been measured by the degenerated four wave mixing (DFWM) technique. In fourth chapter, the structure of low-n Rydberg states doped into supercritical fluids has been investigated in several atomic perturbers. An n-valued Coulombs Force Law, which leads directly to light quanta generating the atomic energy spectrum of hydrogen, has been presented in fifth chapter. Sixth chapter discusses atomic regular polyhedron electronic shell. A new approach on molecular beam depletion has been presented in seventh chapter. Dual conjugate adaptive optics prototype for wide field high resolution retinal imaging has been outlined in eighth chapter. Devices and techniques for sensorless adaptive optics have been described in ninth chapter. A unified approach to analyzing the anisoplanatism of adaptive optical systems has been focused in tenth chapter. Last chapter discusses on renovated quantum physics.

Chapter 1

ATOMIC FORCE MICROSCOPY IN OPTICAL IMAGING AND CHARACTERIZATION

Martin Veis[1] and Roman Antos[2]

[1] Institute of Physics, Faculty of Mathematics and Physics, Charles University
Institute of Biophysics and Informatics, 1st Faculty of Medicine, Charles University

[2] Institute of Physics, Faculty of Mathematics and Physics, Charles University Czech Republic

INTRODUCTION

Atomic force microscopy (AFM) is a state of the art imaging system that uses a sharp probe to scan backwards and forwards over the surface of an object. The probe tip can have atomic dimensions, meaning that AFM can image the surface of an object at near atomic resolution. Two big advantages of AFM compared to other methods (for example scanning tunneling microscopy) are: the samples in AFM measurements do not need to be conducting because the AFM tip responds to interatomic forces, a cumulative effect of all electrons instead of tunneling current, and AFM can operate at much higher distance from the surface (5-15 nm), preventing damage to sensitive surfaces. An exciting and promising area of growth for AFM has been in its combination with optical microscopy. Although the new optical techniques developed in the past few years have begun to push traditional limits, the lateral and axial resolution of optical microscopes are typically limited by the optical elements in the microscope, as well as the Rayleigh diffraction limit of light. In order to investigate the properties of nanostructures, such as shape and size, their chemical composition, molecular structure, as well as their dynamic properties, microscopes with high spatial resolution as well as high spectral and temporal resolving power are required.

Near-field optical microscopy has proved to be a very promising technique, which can be applied to a large variety of problems in physics, chemistry, and biology. Several methods have been presented to merge the optical information of near-field optical microscopy with the measured surface topography. It was shown by (Mertz et al. (1994)) that standard AFM probes can be used for near-field light imaging as an alternative to tapered optical fibers and photomultipliers. It is possible to use the microfabricated piezoresistive AFM cantilevers as miniaturized photosensitive elements and probes. This allows a high lateral resolution of AFM to be combined with near-field optical measurements in a very convenient way. However, to successfully employ AFM techniques into the near-field optical microscopy, several technical difficulties have to be overcome. Artificial periodical nanostructures such as gratings or photonics crystals are promising candidates for new generation of devices in integrated optics.

Precise characterization of their lateral profile is necessary to control the lithography processing. However, the limitation of AFM is that the needle has to be held by a mechanical arm or cantilever. This restricts the access to the sample and prevents the probing of deep channels or any surface that isn't predominantly horizontal. Therefore to overcome these limitations the combination of AFM and optical scatterometry which is a method of determining geometrical (and/or material) parameters of patterned periodic structures by comparing optical measurements with simulations, the least square method and a fitting procedure is used.

AFM PROBES IN NEAR-FIELD OPTICAL MICROSCOPY

In this section we review two experimental approaches of the near-field microscopy that use AFM tips as probing tools. The unique geometrical properties of AFM tips along with the possibility to bring the tip apex close to the sample surface allow optical resolutions of such systems to few tens of nanometers. These resolutions are not reachable by conventional microscopic techniques. For readers who are interested in the complex near-field optical phenomena we kindly recommend the book of (Novotny & Hecht (2006)).

Scattering-type scanning near-field optical microscopy

Scanning near field optical microscopy (SNOM) is a powerful microscopic method with an optical resolution bellow the Rayleigh diffraction limit. The optical microscope can be setup as either an aperture or an apertureless microscope. An aperture SNOM (schematically shown in Figure. 1(a)) uses a metal coated dielectric probe, such as tapered optical fibre, with a submicrometric aperture of diameter d at the apex. For the proper function of such probe it is necessary that d is above the critical cutoff diameter $d_c = 0.6\lambda/n$, otherwise the light propagation becomes evanescent which results in drastic λ dependent loss (Jackson (1975)). This cutoff effect significantly limits the resolution which can be achieved. The maximal resolution is therefore limited by the minimal aperture d $\approx$ λ/10. In the visible region the 50 nm resolution is practically achievable (Hecht (1997)). With the increasing wavelength of illumination light, however, the resolution is decreased.

This leads to the maximum resolution of 1 μm in mid-infrared region, which is non usable for microscopy of nanostructures. To overcome the limitations of aperture SNOM, one can use a different source of near field instead of the small aperture. This source can be a small scatter, such as nanoscopic particle or sharp tip, illuminated by a laser beam. When illuminated, these nanostructures provide an enhancement of optical fields in the proximity of their surface. This is due to a dipole in the tip which is induced by the illumination beam. This dipole itself induces a mirror dipole in the sample when the tip is brought very closely to its surface. Owing to this near-field interaction, complete information about the sample's local optical properties is determined by the elastically scattered light (scattered by the effective dipole emerging from the combination of tip and sample dipoles) which can be detected in the far field using common detectors. This is a basis of the scattering-type scanning near-field optical microscope (s-SNOM). There are two observables of practical importance in the detected signal:

The absolute scattering efficiency and the material contrast (the relative signal change when probing nanostructures made from different materials). The detection of scattered radiation was first demonstrated in the microwave region by (Fee et al. (1989)) (although the radiation was confined in waveguide) and later demonstrated at optical frequencies by using an AFM tip as a scatterer (Zenhausern et al. (1995)) The principle of s-SNOM is shown in Figureure 1(b). Both the optical and mechanical resolutions are determined by the radius of curvature a at the tip's apex and the optical resolution is independent of the wavelength of the illumination beam.

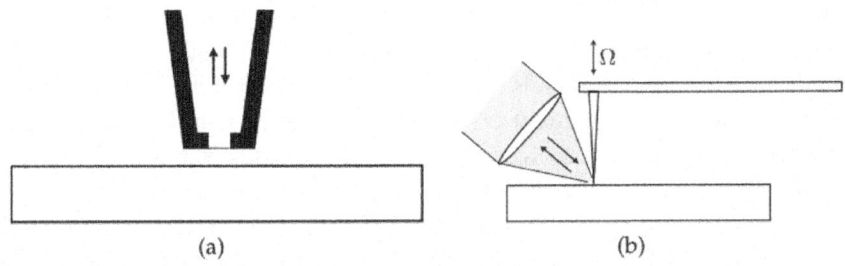

(a)   (b)

Figure. 1: Principles of aperture (a) and apertureless (b) scanning near-field optical microscopies

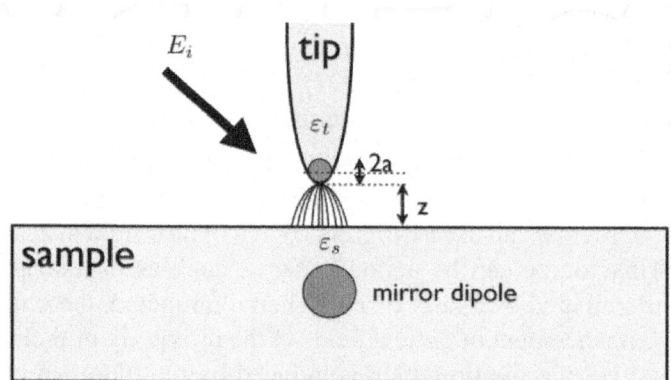

Figure. 2: Schematic view of the simplified theoretical geometry, where the tip was replaced by a small sphere at the tips apex. The sample response is characterized by an induced mirror dipole.

To theoretically solve the complex problem of the realistic scattering of the illuminating light by an elongated tip in the proximity of the sample's surface it is necessary to use advanced electromagnetic theory, which is far beyond the scope of this chapter (readers are kindly referred to the work of (Porto et al. (2000))). However (Knoll & Keilmann (1999b)) demonstrated that the theoretical treatment based on simplified geometry can be used for quantitative calculation of the relative scattering when probing different materials. They have approximated the elongated probe tip by a polarizable sphere with dielectric constant εt , radius a (a $\ll \lambda$) and polarizability (Zayats & Richards (2009))

$$\alpha = 4\pi a^3 \frac{(\varepsilon_t - 1)}{(\varepsilon_t + 2)}.$$

(1)

This simplified geometry is schematically shown in Figureure 2. The dipole is induced by an incident field E_i which is polarized parallel with the tip's axis (z direction). The incident polarization must have the z component. In this case the tip's shaft acts as an antenna resulting in an enhanced near-field (the influence of the incident polarization on the near-field enhancement was investigated by (Knoll & Keilmann (1999a))). This enhanced field exceeds the incident field E_i resulting in the indirect polarization of the sample with dielectric constant ε_s, which fills the half-space z < 0. Direct polarization of the sample by E_i is not assumed. To obtain the polarization induced in the sample, the calculation is approximated by assuming the dipole as a point in the centre of the sphere. Then the near-field interaction between the tip dipole and the sample dipole in the electrostatic approximation can be described by the polarizability $\alpha\beta$ where

$$\beta = \frac{(\varepsilon_s - 1)}{(\varepsilon_s + 1)} \tag{2}$$

Note that the sample dipole is in the direction parallel to those in the tip and the dipole field is decreasing with the third power of the distance. Since the signal measured on the detector is created by the light scattered on the effective sample-tip dipole, it is convenient to describe the near-field interaction by the combined effective polarizability as was done by (Knoll & Keilmann (1999b)). This polarizability can be expressed as

$$\alpha_{\text{eff}} = \frac{\alpha(1 + \beta)}{1 - \frac{\alpha\beta}{16\pi(a+z)^3}}, \tag{3}$$

where z is the gap width between the tip and the sample. For a small particle, the scattered field amplitude is proportional to the polarizability (Keilmann & Hillenbrand (2004))

$$E_s \propto \alpha_{\text{eff}} E_i. \tag{4}$$

Since the quantities ε, β and α are complex, the effective polarizability can be generally characterized by a relative amplitude s and phase shift ϕ between the incident and the scattered light

$$\alpha_{\text{eff}} = se^{i\varphi}. \tag{5}$$

The validity of the theoretical approach described above was determined by numerous s-SNOM studies published by (Hillenbrand & Keilmann (2002); Knoll & Keilmann (1999b); Ocelic & Hillenbrand (2004)). Good agreement between experimental and theoretical s-SNOM contrast was achieved

Recalling the Equation (3) it is important to note that the change of the illumination wavelength will lead to changes in the scattering efficiency as the values of the dielectric constants ε_s and ε_t will follow dispersion relations of related materials. This allows to distinguish between different materials if the tip's response is flat in the spectral region of interest. Therefore the proper choice of the tip is important to enhance the material contrast and the resolution. (Cvitkovic et al. (2007)) reformulated the coupled dipole problem and derived the formula for the scattered amplitude in slightly different form

$$E_s = (1 + r)^2 \frac{\alpha(1 + \beta)}{1 - \frac{\alpha\beta}{16\pi(a+z)^3}}. \tag{6}$$

They introduced Fresnel reflection coefficient of the flat sample surface. This is important to account for the extra illumination of the probe via reflection from the sample which was neglected in Equations (3) and (4). The detected signal in s-SNOM is a mixture of the near-field scattering and the background scattering from the tip and the sample. Prior to the description of various experimental s-SNOM setups it is important to note how to eliminate the unwanted background scattering from the detector signal. For this purpose we have calculated the distance dependence of α_{eff}. The result is displayed in Figureure 3. As one can see from this Figureure, the scattering is almost constant for distances larger than 2a. On the other hand for very short distances (very closely to the sample) both the scattering amplitude and the scattering phase drastically increase.

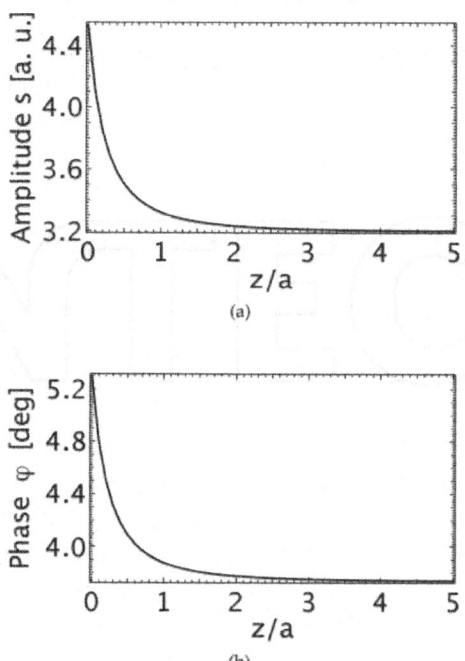

(a)

(b)

Figure. 3: Theoretically calculated dependence of the near-field scattering amplitude s (a) and phase ϕ (b) on the tip-sample distance z.

This occurs for various materials with various dielectric constants demonstrating the near-field interaction. When the tip is illuminated by a focused laser beam, only a small portion of the incident light reaches the gap between the tip and the sample and contributes to the near-field. Therefore the detected signal is mainly created by the background scattering. The nonlinear behavior of the $\alpha_{eff}(z)$ is employed to filter out the unwanted background scattering which dominates in the detected signal. This can be done if one employs tapping mode with a tapping frequency Ω into the experimental setup.

The tapping of amplitude $\Delta z \approx a \approx 20nm$ modulates the near-field scattering much stronger than the background scattering. The nonlinear dependence of $\alpha_{eff}(z)$ will introduce higher harmonics in the detected signal. The full elimination of the background is done by demodulating the detector signal at the second or higher harmonic of Ω as was demonstrated by (Hillenbrand & Keilmann (2000)) and others. There are various modifications of the s-SNOM experimental setup. Schematic views of interferometric s-SNOM experimental setups with heterodyne, homodyne and pseudohomodyne detection are displayed in Figure 4.

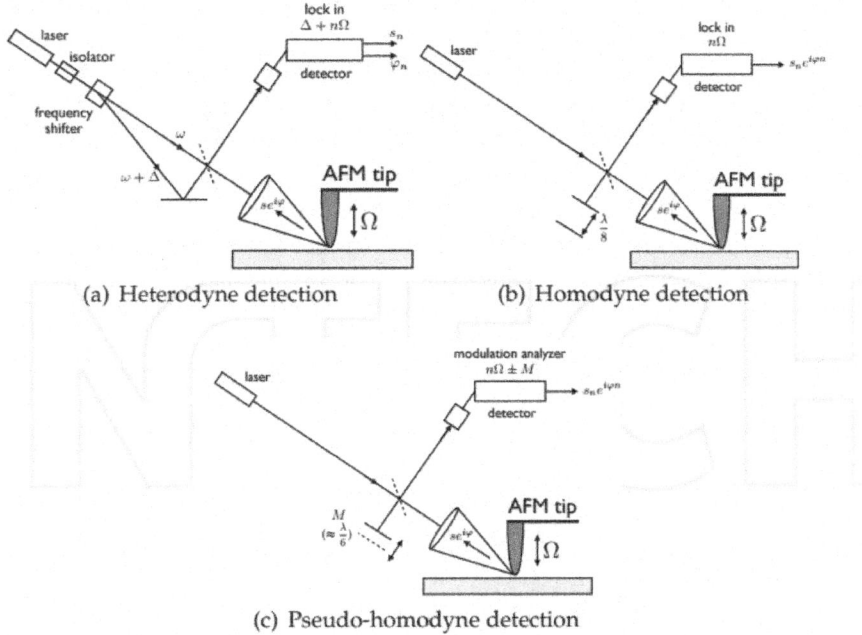

(a) Heterodyne detection   (b) Homodyne detection

(c) Pseudo-homodyne detection

Figure. 4: Schematic views of experimental interferometric s-SNOM setups

The heterodyne detection system developed by (Hillenbrand & Keilmann (2000)) uses a HeNe laser with output power of $\approx 1mW$ as the illumination source. The beam passes through the optical isolator to filter the back reflections from the frequency shifter. The frequency shifter creates a reference beam with the frequency shifted by $\Delta = 80MHz$ which interferes with the backscattered light from the sample in a heterodyne interferometer. The detected intensity is therefore $I = I_{ref} + I_s + 2\sqrt{I_{ref}I_s}\cos(\Delta t + \varphi).$ The signal is processed in high frequency lock-in amplifier which operates on the sum frequency $\Delta + n\Omega$. Here n is the number of higher harmonic. The lock-in amplifier gives two output signals. One is proportional to scattering amplitude while the second is proportional to the phase of the detector modulation at frequency $\Delta + n\Omega$. When the order of harmonic n is sufficiently large the signal on the lock-in amplifier is proportional to sn and ϕn. This means that using higher harmonics, one can measure pure near-field response directly. Moreover such experimental setup has optimized signal/noise ratio.

The influence of higher harmonic demodulation on the background filtering is demonstrated in Figure 5. In this Figure the tip was used to investigate gold islands on Si substrate. For n = 1 the interference of different background contributions is clearly visible for z > a. Such interference may overlap with the important near-field interaction increase for z < a which leads to a decrease of the contrast. Taking into account the second harmonic (n = 2) one can see a rapid decrease of the interference which allow near-field interactions to be more visible. For the third harmonic (n = 3) the near-field interaction becomes even steeper. Because the tip is periodically touching the sample, a non sinusiodal distortion of the taping motion can be created by the mechanical motion. This leads to artifacts in the final microscopic image which are caused by the fact that the higher harmonics $n\Omega$ are excited also by mechanical motion.

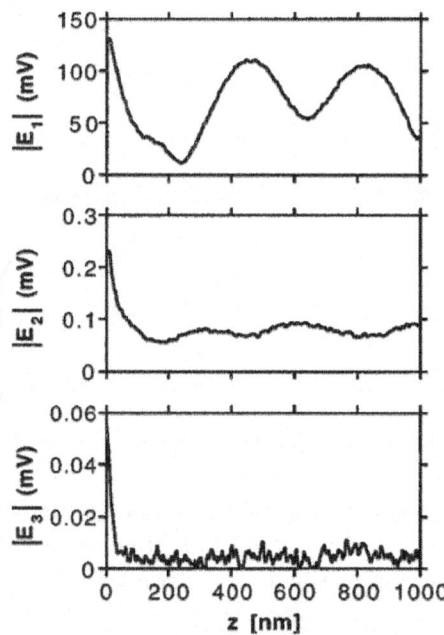

Figure. 5: Optical signal amplitude |En| vs distance z between tip and Au sample, for different harmonic demodulation orders n. ©2002 American Institute of Physics (Hillenbrand & Keilmann (2002))

These mechanical harmonics cause direct modulations of the optical signals resulting in a distorted image. (Hillenbrand et al. (2000)) demonstrated that the artifacts depend on the sample and the tapping characteristics (such as amplitude, etc.). They have found that these mechanical artifacts are negligible for small $\Delta z < 50$nm and large set points $\Delta z / \Delta z_{free} > 0.9$.

In the mid-infrared region the appropriate illumination source was a $CO2$ laser owing to its tunable properties from 9.2 to 11.2 µm. The attenuated laser beam of the power ≈ 10mW was focused by a Schwarzschild mirror objective (NA = 0.55) to the tip's apex. The polarization of the incident beam was, as in the previous case, optimized to have a large component in the direction of the tip shaft. This lead to a large enhancement of the near field interaction and increased the image contrast. The incident laser beam was split to create a reference which was reflected on a piezoelectrically controlled moveable mirror (Figure 4(b)). This mirror and the scattering tip created a Michelson interferometer.

Using a homodyne detection the experimental setup was continuously switching the mirror between two positions. The first position corresponded to the maximum signal of the n-th harmonic at the lock-in amplifier (positive interference between the near-field scattered light and the reference beam) while the second position was moved by a $\lambda/8$ (90° shift of reference beam).

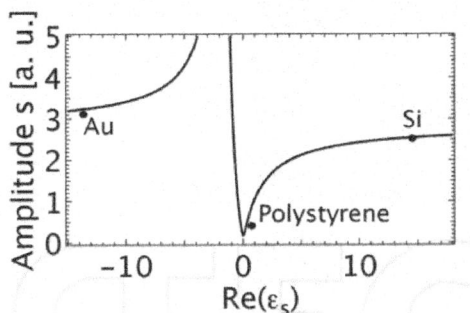

Figure. 6: Theoretically calculated the near-field scattering amplitude s as a function of the real part of ε_s. The imaginary part of ε_s was set to 0.1 and the tip was considered as P_t.

With the experimental setup the detection of the amplitude and the phase of the near-field scattering was possible to detect, obtaining the near-field phase and amplitude contrast images. Further improvement of the background suppression was demonstrated by (Ocelic et al. (2006)) using a slightly modified homodyne detection with a sinusoidal phase modulation of the reference beam at frequency M (see Figure 4(c)). This lead to the complete reduction of the background interference. As we already mentioned and as is clearly visible from the equations described above, the near-field scattering depends on the dielectric function of the tip and the sample. We have calculated the amplitude of the near-field scattering as a function of the real part of ε_s using Equation (3). The tip is assumed to be P_t ($\varepsilon_t = -5.2 + 16.7i$) and the sphere diameter $a = 20$nm. The result is depicted in Figure 6. The imaginary part of ε_s was set to 0.1. The inserted dots represent the data for different materials at illumination wavelength $\lambda = 633$nm (Hillenbrand & Keilmann (2002)). As can be clearly seen from Figure 6, owing to different scattering amplitudes, a good contrast in the image of nanostructures consists of Au, Polystyrene and Si components should allow for easily observable images. Indeed, this was observed by (Hillenbrand & Keilmann (2002)) and is shown in Figure 7 . The AFM topography image itself can not distinguish between different materials. However, due to the material contrast, it is possible to observe different material structures in the s-SNOM image. This is consistent with the theoretical calculation in Figure 6. The lateral resolution of the s-SNOM image in Figure 7 is 10 nm.

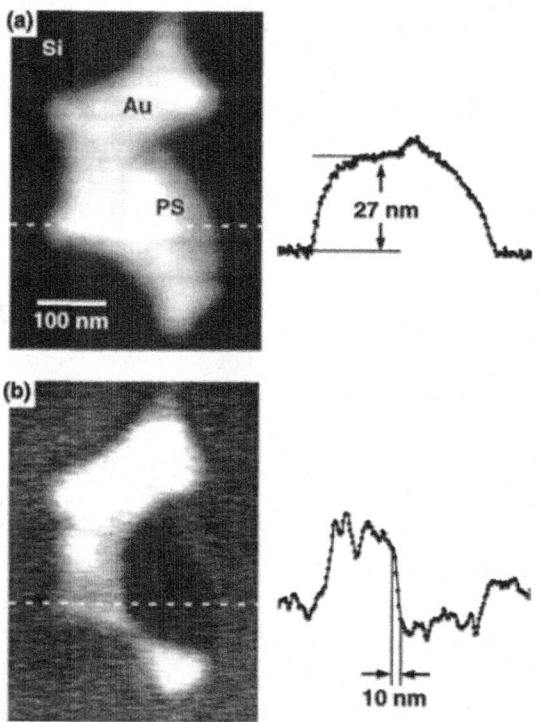

Figure. 7: Auisland on Si observed in (a) topography, (b) optical amplitude |E3|, with adjoining polystyrene particle.

Tip enhanced fluorescence microscopy Owing to its sensitivity to single molecules and biochemical compositions, fluorescence microscopy is a powerful method for studying biological systems. There are various experimental setups of fluorescent microscopes that exceed the Rayleigh diffraction criterion which limits the practical spatial resolution to ≈ 250nm. Recent modifications of conventional confocal microscopy, such as $4 - \pi$ (Hell & Stelzer (1992)) or stimulated emission depletion (Klar et al. (2001)) microscopies have pushed the resolution to tens of nanometers. Although these techniques offer a major improvement in the field of fluorescent microscopy, they require high power laser beams, specially prepared fluorophores and provide slow performance (not suitable for biological dynamics).

The line scans give evidence of purely optical contrast at 10 nm resolution, and of distinct near-field contrast levels for the three materials. ©2002 American Institute of Physics (Hillenbrand & Keilmann (2002)) Experimental setups of s-SNOM, as described in detail above, can be modified to sense fluorescence from nanoscale structures offering an alternative method to confocal microscopy. The near-field interaction between the sample and the tip causes the local increase of the one-photon fluorescence-excitation rate. The fluorescence is then detected by a single-photon sensitive avalanche photodiode. Such an experimental technique is called tip-enhanced fluorescence microscopy (TEFM) and its setup is schematically shown in Figure 8.

There are two physical effects detected. The first one is an increase of detected fluorescence signal due to the near-field enhancement. The second one is the signal decrease due to the fluorescence quenching. These effects were demonstrated by various authors, for example by (Anger et al. (2006)). The fluorescence enhancement is proportional to the real part of the dielectric constant of the tip. On the other hand the fluorescence quenching is proportional to the imaginary part of the same dielectric function. Since these effects manifest themselves at short distances (bellow 20 nm), they can be used to obtain nanoscale resolution. Because the fluorescence enhancement leads to higher image contrast, silicon AFM tips are often used (due to their material parameters) for fluorescence studies of dense molecular systems. In TEFM the illumination beam stimulates simultaneously a far-field fluorescence component S_{ff}, which is coming from direct excitation of fluorophores within the laser focus, and a near-field component S_{nf}, which is exited by a near-field enhancement. One can then define

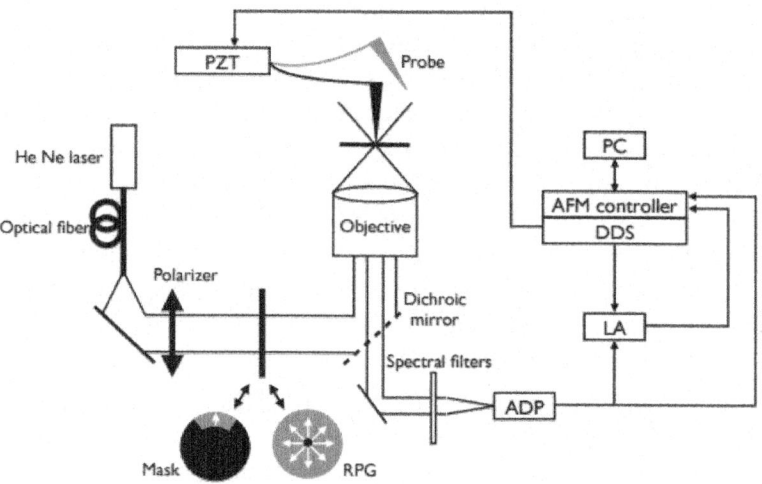

Figure. 8: Experimental setup of TEFM. RPG - radial polarization generator; PZT-piezoelectric transducer; ADP - avalanche photodiode; LA - lock-in amplifier; DDS - digital synthesizer.

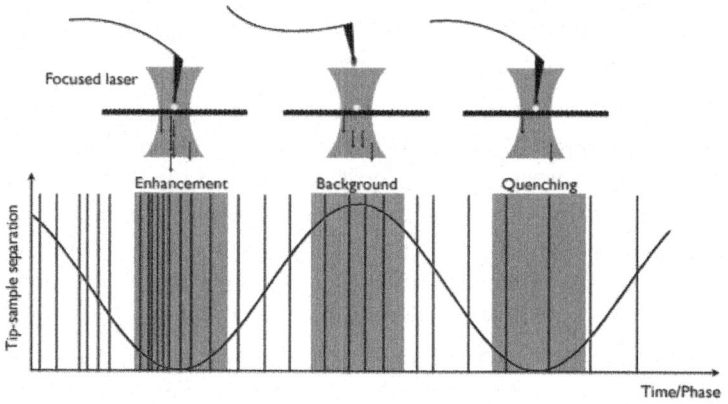

Figure. 9: Schematic picture of fluorescence modulation by AFM tip oscilation.

a contrast (C) of TEFM as

$$C = \frac{S_{nf}}{S_{ff}}.$$

(7)

Similarly to the s-SNOM setup it is possible to enhance the contrast and resolution of TEFM using a tapping mode in AFM and a demodulation algorithm for detected signal. Such process can be done by lock-in amplification. The scheme of the fluorescence modulation by an AFM tip oscillation is shown in Figure 9. When the tip is in the highest position above the sample no near-field interaction occurs. The detected signal is therefore coming from the background scattering excitation. If the tip is approaching the sample the fluorescence rate becomes maximally modified. The detected signal is either positive or negative depending on the fluorescence enhancement or quenching. TEFM example images of high density CdSe/ZnS quantum dots are shown in Figure 10. An improvement of the lateral resolution and contrast is clearly visible when using a TEFM with lock-in demodulation detection. The resolution of 10 nm, which is bellow the resolution of other fluorescence microscopies, demonstrates the main advantage of TEFM systems.

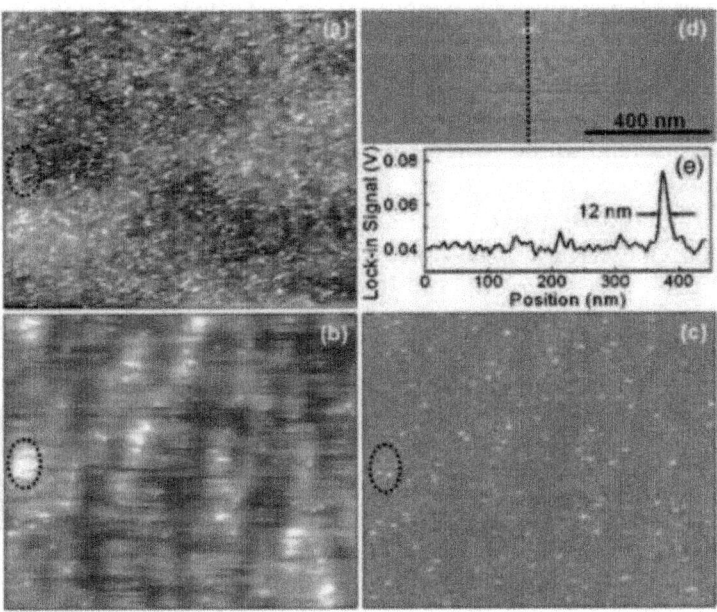

Figure. 10: High-resolution images of quantum dots. (a) AFM topography image; (b) photon-sum image; (c) TEFM image using lock-in demodulation. (a)–(c) are for a 5x5 μm2 field of view. (d) TEFM image of a single quantum dot; (e) signal profile specified by the dotted line in (d). ©2006 American Institute of Physics (Xie et al. (2006))

AFM VERSUS SCATTEROMETRY

Recent advances of integrated circuits, including shortened dimensions, higher precision, and more complex shapes of geometric features patterned by modern lithographic methods, also requires higher precision of characterization techniques. This section briefly reviews some improvements of AFM and optical scatterometry, their comparison (with mutual advantages and disadvantages), and their possible cooperation in characterizing the quality of patterned nanostructures, especially in determining critical dimensions (CD), pattern shapes, line width roughness (LWR), or line edge roughness (LER).

AFM in the critical dimension (CD) mode with flared tips

It has been frequently demonstrated that accurate monitoring of sidewall features of patterned lines (or dots or holes) by AFM requires probe tips with special shapes and post processing algorithms to remove those shapes from the acquired images of the patterned profiles. A conventional AFM tip (with a conical, cylindrical, or intermediate shape) even with an infinitely small apex is only capable of detecting the surface roughness on horizontal surfaces (on the top of patterned elements or on the bottom of patterned grooves), but cannot precisely detect sidewall angles, LER, or particular fine sidewall features, as depicted in Figure. 11(a). Here the oscillation of the probe while scanning is only in the vertical direction so that the inverse profile of the tip is obtained instead of the correct sidewall shape. Only patterns with sidewall slopes smaller than the tip slopes (e.g., sinusoidal gratings) can be accurately detected after applying an appropriate image reconstruction transform such as the one shown by (Keller (1991)).

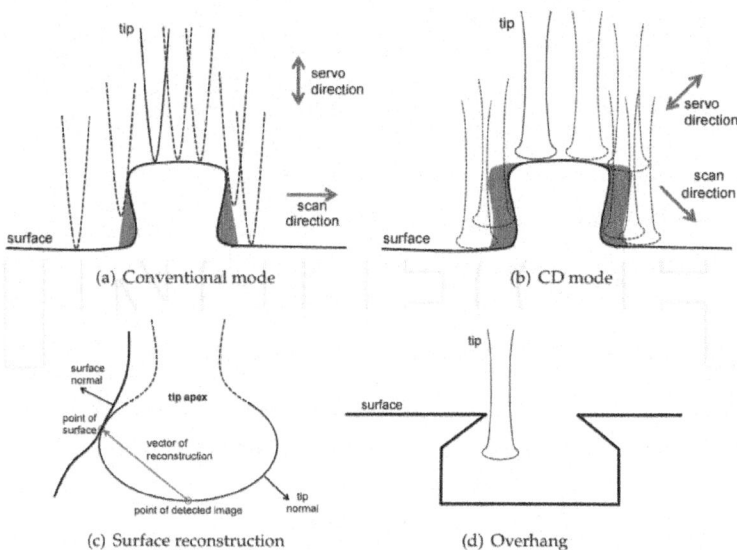

Figure. 11: AFM scanning of a patterned element in the conventional (a) and CD (b) modes with a conventional (a) and flared (b) tip apex. Postprocessing reconstruction of the surface uses the vector of reconstruction (c). The method can fail on overhang structures (d).

On the other hand, a CD tip, fabricated with a flared apex radius such as in (Liu et al. (2005)), can provide an accurate 3D patterned profile provided that it is applied in the CD mode. This mode, unlike the conventional deep trench mode where the tip only oscillates in the vertical direction, requires the tip to oscillate in both vertical and horizontal directions to follow the full surface topography for which multiple vertical points are possible for the same horizontal position, as depicted in Figure. 11(b). Analogously to conventional AFM scanning with a nonideal tip with a finite apex, the CD AFM scan also requires a post processing reconstruction of the real surface profile. An example of such a reconstruction, demonstrated by (Dahlen et al. (2005)), is the application of the reconstruction vector utilizing the fact that the normal to the surface is identical with the normal to the tip at each contact point, as displayed in Figure. 11(c). The method also utilizes algorithms of "reentrant" surface description. The CD-AFM scanning can obviously fail for highly undercut surfaces for which the tip apex is not sufficiently flared, as visible in Figure. 11(d). However, such a structure can be advantageously used to carry out a topography measurement of the actual tip sidewall profile, as also described by (Dahlen et al. (2005)). Such a structure, designed solely for this reason, is called an overhang characterizing structure. The advantages of the CD-AFM are that it is a nondestructive method (unlike cross-sectional SEM) which provides a direct image of the cross-sectional surface profile with relatively high precision. However, the image becomes truly direct after an appropriate post processing procedure removing the tip influence. Moreover, some profile features cannot be revealed such as the precise shape of the top sharp corners, the exact vertical positions and radii of fine sidewall features, and—most importantly—the real shape of sharp bottom corners of the grooves.

AFM used for line edge roughness (LER) characterization

As the dimensions of patterned structures shorten to the nanometer scale, the LER and the LWR become important characteristics. Following (Thiault et al. (2005)), we briefly define the LER and LWR as follows:

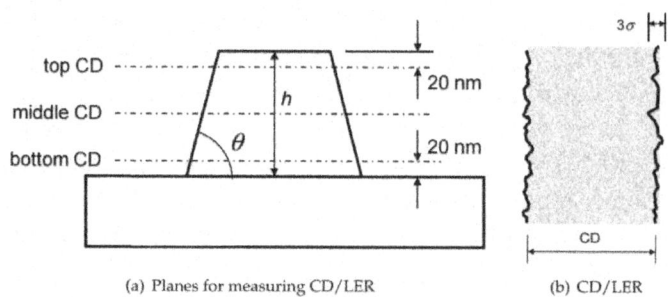

(a) Planes for measuring CD/LER   (b) CD/LER

Figure. 12: Definitions of measuring CD and LER in different planes for AFM in the CD mode.

The LWR is defined as three standard deviations (denoted 3σ) of the scanned linewidth variations at a height determined by the AFM user, while the CD represents the averaged value of the scanned dimension. Analogously, the LER is defined as its one-edge version, measured as three standard deviations of the variations from the straight line edge. Unfortunately, the CD-AFM scanning cannot accurately detect the linewidth at the very top and—especially—at the very bottom of the patterned grooves (which is due to the finite size of the tip apex). For this reason it is usually determined at some small height from the top (typically 20 nm, determined by the tip used), at the middle, and at some small height above the bottom (hence the top, middle, and bottom CD and LWR/LER, respectively). The corresponding planes are depicted in Figure. 12(a). The geometries for measuring the CD and the LER (3σ) at a chosen height are depicted in Figure. 12(b). Although the LWR and LER at specified heights are also calculated from direct CD-AFM images, they can be affected by some further defects. As an example, consider a line whose both edges have equal LER values which are mutually uncorrelated. According to the statistical theory, $\sigma_{LWR}^2 = 2\sigma_{LER}^2$ should be valid, so that the LWR should be $2^{1/2}$ times higher than the LER. However, (Thiault et al. (2005)) have shown that the stage drift (breaking the relative position of the tip and sample) during long-time measurement affect the LER considerably more than the LWR, because the time between the detection of two adjacent edges is much shorter than the time between two sequential scans of the same edge.

Critical dimensions measured by scatterometry

Optical scatterometry, most often based on spectroscopic ellipsometry and sometimes combined with spectrophotometry (light intensity reflectance or transmittance), is an optical investigation method which combines optical measurements (typically in a wide spectral range, utilizing visible light with near ultraviolet and infrared edges) together with rigorous optical calculations. The spectra are calculated with varied input geometrical parameters of the patterned structure (optical CDs) and compared to the measured values to minimize the difference (optical error) as much as possible, employing the least square method for the optical error. The algorithm is usually referred to as the optical fitting procedure, and the obtained optical CDs are referred to as the optically fitted dimensions. Various authors have presented the use of specular (0th-order diffracted) spectroscopic scatterometry to determine linewidths, periods, depths, and other fine profile features not accessible by AFM (such as the above mentioned bottom corners of grooves), as shown by (Huang & Terry Jr. (2004)). Obvious advantages of scatterometry is (besides non-destructiveness) higher sensitivity and no contact with any mechanical tool. Simply speaking, a photon examines the structure as it really stands and gives the true answer.

Another advantage is the possibility to integrate an optical apparatus (most often a spectral ellipsometer) into a lithographer or deposition apparatus for the in situ monitoring of deposition and lithographic processes. On the other hand, scatterometry has some disadvantages: Spectral measurements are indirect, and the measured spectra sometimes require very difficult analyses to reveal the real profile of the structure, which should be approximately known before starting the fitting procedure (to use it as an initial value). Moreover, the spectra can contain too many unknown parameters, or at least some vaguely known parameters. As an example, consider a grating made as periodic wires patterned from a Ta film deposited on a quartz substrate, which was optically investigated by (Antos et al. (2006)). The unknown parameters in the beginning were not only the geometrical dimensions and shape of the wires, but also the material properties of the Ta film, which were altogether correlated. Therefore, the first analysis was performed on a nonpatterned reference Ta film to determine the refractive index and extinction coefficient of Ta, as well as the thickness of the native oxide overlayer of Ta_2O_5.

The obtained parameters were used as known constants within the second analysis, which was made on the Ta wire grating. This second analysis yielded the values of period, depth, linewidths, and the sidewall shape of the Ta wires. The sidewall shape was analytically approximated as paraboloidal, determined by two parameters: the top (smallest) linewidth (same as the bottom linewidth) and the middle (maximum) linewidth. Although the obtained geometry was revealed with higher precision than the geometry obtained by a direct method (unpublished SEM images in this case), all the obtained parameters were affected by a slight difference from the assumed paraboloidal sidewalls and by a native Ta_2O_5 overlayers that developed on the sidewalls (which were not taken into account in simulations). Each such difference or negligence from the real sample can contribute to discrepancy between the optical CDs and the real dimensions. Simply speaking, the advantage of high sensitivity can easily become a disadvantage, when the optical con Figureuration is too sensitive to undesired features. To illustrate the basic difference between conventional AFM and scatterometric measurements, consider a shallow rectangular grating patterned on the top of a 32-nm-thick

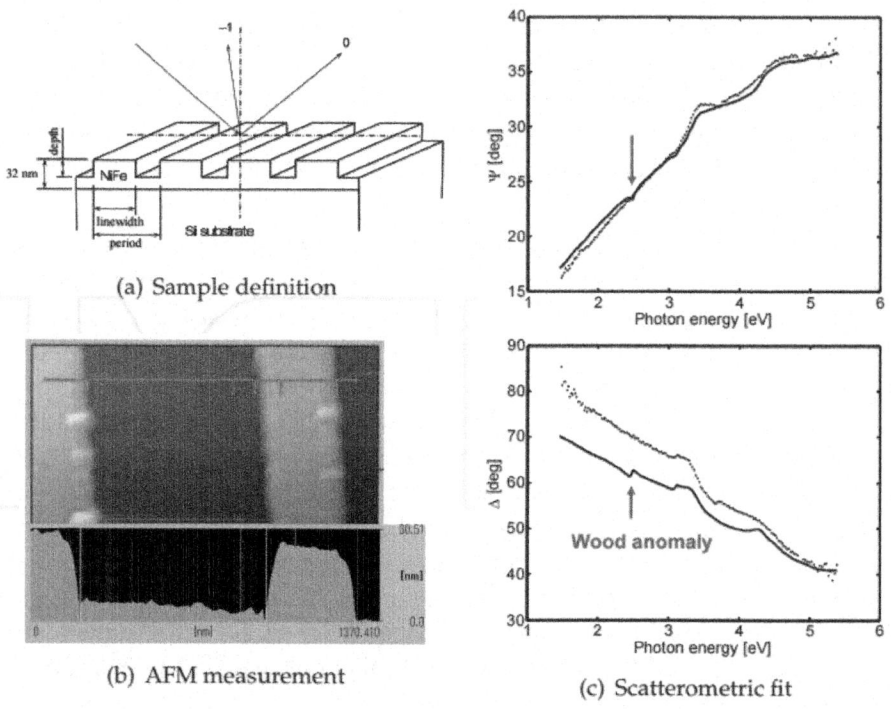

(a) Sample definition

(b) AFM measurement

(c) Scatterometric fit

Figure. 13: Investigation of a sample (a) by AFM (b) and scatterometry (c).

Permalloy (NiFe) film deposited on a Si substrate, with geometry depicted in Figure. 13(a) and with more details in (Antos et al. (2005d)). The comparison of nominal geometrical parameters (those intended by the grating manufacturer) with parameters determined by AFM [Figure. 13(b)] and scatterometry [based on spectroscopic ellipsometry performed at three angles of incidence, 60, 70, and 80∘ , the last of which is displayed in Figure. 13(c)] is listed in Table 1.

Table 1: Comparison of grating geometrical parameters obtained by AFM and scatterometry, together with nominal ones (intended by the manufacturer)

Parameter	nominal	AFM	scaled AFM	scatterometry
period	1000	1091.5	1000*	1000**
linewidth	500	359.4	329.3	307.2
NiFe thickness	32	—	—	—
relief depth	16	21.7	21.7	24.3

*fixed value
**just verified from the position of the Wood anomaly

Here the AFM measurement provides a direct image of the grating's relief profile (for our purposes to scan a shallow relief the conventional-mode AFM is appropriate), but the horizontal values (period and line width) are affected by a wrong-scale error (9 %, which is quite high). According to our experiences, the period of patterns made by lithographic processes is always achieved with high precision, so that we scale the horizontal AFM parameters to obtain the nominal 1000 nm period and to keep the same period-to-line width ratio (the verical depth is kept without change). The scaled AFM linewidth (329.3 nm, measured at the bottom) now corresponds well to the linewidth determined by scatterometry (307.2 nm); the 22 nm difference is probably due to the finite size of the apex of the AFM tip, which was previously explained in Figure. 11(a). As visible in Figure. 13(c), the ellipsometric spectrum is not sensitive to the grating period, which is due to the shallow relief. For this reason the grating period could not be included in the fitting procedure. However, the period can be easily and with high precision verified by observing the spectral positions of Wood anomalies, which depend only on the period and the angle of incidence. In our case we observe the −1st-order Rayleigh wavelength (wavelength at which the −1st diffraction order becomes an evanescent wave) which provides precise verification of the nominal period of 1000 nm.

The 2.6 nm difference between the AFM and scatterometric values of the relief depth is difficult to explain. It might be again due the wrong vertical scale of the AFM measurement, or it might be due to some negligences of the used optical model such as the native top NiFe oxide overlayer or inaccurate NiFe optical parameters. Nevertheless, AFM and scatterometry differ from each other much less than how they differ from the nominal value, which indicates their adequacy. Finally, none of the methods were able to determine the full thickness of the deposited NiFe film (whose nominal value is 32 nm). While AFM was obviously disqualified in principle because it only detects the surface, optical scatterometry could provide this information if a reference sample (nonpatterned NiFe film simultaneously deposited on a transparent substrate) were investigated by energy transmittance measurement (again spectrally resolved for higher precision). Another way to improve accuracy or the number of parameters to be resolved is to include higher diffraction orders in the analysis or to measure additional spectra such as magneto-optical spectroscopy (utilizing the magneto-optical anisotropy of NiFe).

As an example, consider a grating made of Cr(2 nm)/NiFe(10 nm) periodic wires deposited on the top of a Si substrate. (Antos et al. (2005b)) used magneto-optical Kerr effect spectroscopy combining the analysis of the 0th and −1st diffraction order to determine the thicknesses of native oxide overlayers on the top of the Cr capping layer and on the top of Si substrate. It was shown that the 0th order of diffraction is more sensitive to features present both on wires and between them, whereas higher orders are more sensitive to differences between wires and grooves.

LINE EDGE ROUGHNESS DETERMINED BY SCATTEROMETRY

Besides measuring CDs, (Antos et al. (2005a)) have also shown that scatterometry is capable of evaluating the quality of patterning with respect to LER. Consider a pair of gratings similar to the one previously described, i.e., Cr(2 nm)/NiFe(10 nm) wires on the top of a Si substrate. AFM measurements of the two samples are shown in Figure. 14, where Sample 2 has obviously higher LER than Sample 1. It is well known that p-polarized light is considerably more sensitive to surface features than s-polarized light. For this reason, magneto-optical Kerr effect spectroscopy in the −1st diffraction order was analyzed in a conFigureuration where r_{pp} (amplitude reflectance for the p-polarization) is close to zero. Since the Kerr rotation and ellipticity are approximately equal to the real and imaginary components of the complex ratio r_{sp}/r_{pp} (here for the −1st order), they are very sensitive to the the wire edges. Figure. 15(a) displays experimental spectra measured on Sample 2 compared with two different models.

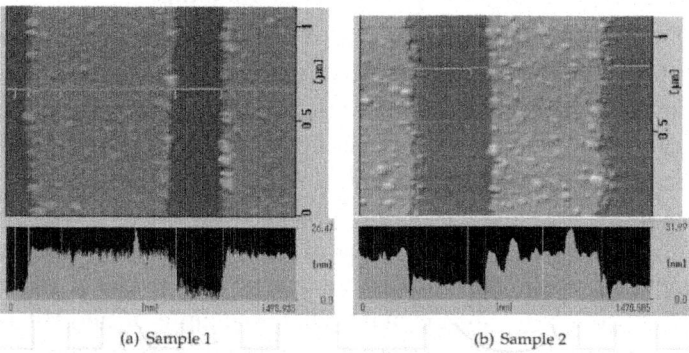

(a) Sample 1   (b) Sample 2

Figure. 14: AFM measurements of two Cr/NiFe wire samples with different LER.

First, the rigorous couple wave analysis (RCWA) is a rigorous method assuming diffraction on a perfect grating with ideal edges. Second, the local mode method (LMM) locally treats the grating structure as a uniform multilayer, neglecting thus the optical effect of the wire edges. Since none of the models corresponds well to the measured values, the reality is somewhere in the middle. To include the effect of LER, we define a third model (optical LER method) as follows:

$$r_{pp}^{LER} = r_{pp}^{LMM} + \eta(r_{pp}^{RCWA} - r_{pp}^{LMM})$$

(8)

Where r_{pp}^{LMM} and r_{pp}^{RCWA} are reflectances calculated by the LMM and RCWA methods, and η is a parameter whose values can be between zero and one. For the case $\eta = 1$ the sample behaves as a perfect grating with ideal edges $(r_{pp}^{LER} = r_{pp}^{RCWA})$, so that LER is zero. For the opposite case $\eta = 0$ the grating behaves as a random formation of islands with the wire structure, with their relative area equal to the grating filling factor, so that LER is infinite or at least higher than the line width. In reality the η parameter will be somewhere in the middle and thus will provide the desired information about the LER. A fitting procedure carried on the two samples from Figure. 14 revealed the values of η = 0.70 for Sample 1 and η = 0.53 for Sample 2, which corresponds well to the obvious quality of the samples. The fitted spectra of the optical LER method are displayed in Figure. 15(b)

Joint Afm-Scatterometry Method

From the above comparisons, the mutual advantages and disadvantages of both AFM and scatterometry are obvious. For complex structures for which none of them can reveal all desired parameters when used solely, it is better to use them both. For instance, AFM can be used as the first method to determine the pattern shape and as many CDs as possible. Then, the obtained CDs are used as initial values for the optical fitting procedure in the frame of scatterometry, or some AFM parameters can be fixed (such as depth and top line width) and the remaining parameters can be fitted by scatterometry.

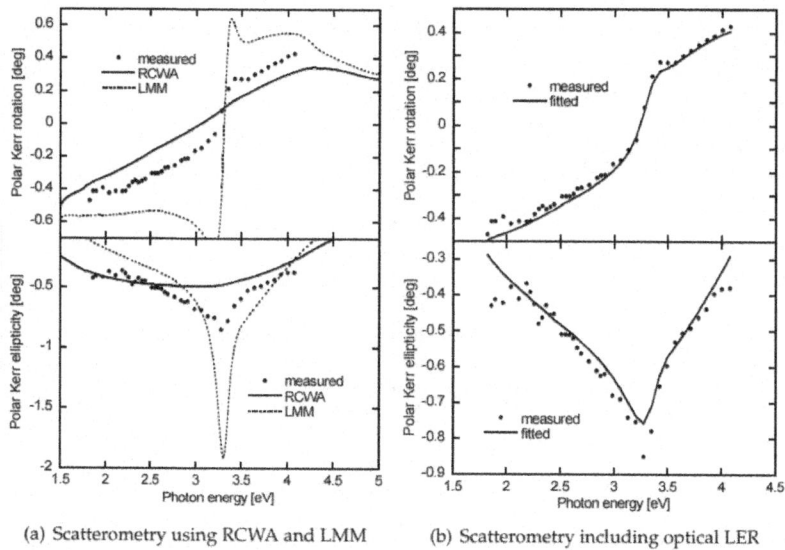

(a) Scatterometry using RCWA and LMM   (b) Scatterometry including optical LER

Figure. 15: Experiment and modeling of MO spectroscopy in the −1 st diffraction order with p-polarized incident light assuming a perfect grating (a) and including optical LER (b).

As an example, consider a nearly-sinusoidal surface-relief grating patterned on the top of a thick epoxy layer with a refractive index close to the index of glass on which the epoxy was deposited. A sample of such a structure was investigated by (Antos et al. (2005c)) to obtain the following results: First, a detailed AFM scan of the surface provided the precise shape of the relief function, being something between a sinusoidal and a triangular function,

$$f(x) = a\frac{d}{2}\sin\frac{2\pi x}{\Lambda} + (1-a)\frac{2d}{\Lambda}x,$$

(9)

where a is a parameter of sharpness (the ratio between the sinusoidal and triangular shape), d is the depth of the grating, and Λ is its period. The AFM scan thus determined the period $\Lambda = 9365$ nm, depth of about d = 700 nm, and the parameter of sharpness a = 0.6. The period and the parameter of sharpness were then fixed as constants and used in a spectroscopic ellipsometry investigation to find more precisely the depth d = 620 nm and the values Δn of how much the epoxy's refractive index differs from the index of the glass substrate.

CONCLUSION

In this chapter we have shown that AFM tips can be used effectively as near-field probes in near-field microscopy. Using a proper experimental setup one can resolve nanostructures down to 10 nm indenpendently of the illumination wavelength, which can be chosen between visible and infrared region.

AFM and optical applications were also reviewed with respect to measuring geometries and dimensions of laterally patterned nanostructures. Both AFM in the CD mode and scatterometry were capable of providing valuable information on periodic relief profiles with different mutual advantages and disadvantages.

ACKNOWLEDGEMENT

This work is part of the research plan MSM 0021620834 financed by the Ministry of Education of the Czech Republic and was supported by the Grant Agency of the Czech Republic (no. P204/10/P346 and 202/09/P355) and a Marie Curie International Reintegration Grant (no. 224944) within the 7th European Community Framework Programme.

REFERENCES

1. Anger, P., Bahradwaj, P. & Novotny, L. (2006). Enhancement and quenching of single-molecule fluorescence, Phys. Rev. Lett. Vol. 96(No. 11): 113002.

2. Antos, R., Mistrik, J., Yamaguchi, T., Visnovsky, S., Demokritov, S. O. & Hillebrands, B. (2005a). Evaluation of the quality of Permalloy gratings by diffracted magneto-optical spectroscopy, Opt. Express Vol. 13(No. 12): 4651–4656.

3. Antos, R., Mistrik, J., Yamaguchi, T., Visnovsky, S., Demokritov, S. O. & Hillebrands, B. (2005b). Evidence of native oxides on the capping and substrate of permalloy gratings by magneto-optical spectroscopy in the zeroth- and first-diffraction orders, Appl. Phys. Lett. Vol. 86(No. 23): 231101.

4. Antos, R., Ohlidal, I., Franta, D., Klapetek, P., Mistrik, J., Yamaguchi, T. & Visnovsky, S. (2005). Spectroscopic ellipsometry on sinusoidal surface-relief gratings, Appl. Surf. Sci. Vol. 244(No. 1-4): 221–224.

5. Antos, R., Pistora, J., Mistrik, J., Yamaguchi, T., Yamaguchi, S., Horie, M., Visnovsky, S. & Otani, Y. (2006). Convergence properties of critical dimension measurements by spectroscopic ellipsometry on gratings made of various materials, J. Appl. Phys. Vol. 100(No. 5): 054906.

6. Antos, R., Veis, M., Liskova, E., Aoyama, M., Hamrle, J., Kimura, T., Gustafik, P., Horie, M., Mistrik, J., Yamaguchi, T., Visnovsky, S. & Okamoto, N. (2005). Optical metrology of patterned magnetic structures: deep versus shallow gratings, Proc. SPIE Vol. 5752(No. 1-3): 1050–1059.

7. Cvitkovic, A., Ocelic, N. & Hillenbrand, R. (2007). Analytical model for quantitative prediction of material contrasts in scattering-type near-field optical microscopy, Opt. Express Vol. 15(No. 14): 8550.

8. Dahlen, G., Osborn, M., Okulan, N., Foreman, W., Chand, A. & Foucher, J. (2005). Tip characterization and surface reconstruction of complex structures with critical dimension atomic force microscopy, J. Vac. Sci. Technol. B Vol. 23(No. 6): 2297–2303.

9. Fee, M., Chu, S. & Hänsch, T. W. (1989). Scanning electromagnetic transmission line microscope with sub-wavelength resolution, Optics Communications Vol. 69(No. 3-4): 219–224. Hecht, B. (1997).

10. Facts and artifacts in near-field microscopy, J. Appl. Phys. Vol. 81(No. 6): 2492–2498.

11. Hell, S. W. & Stelzer, E. H. K. (1992). Fundamental improvement of resolution with a 4Pi-confocal fluorescence microscope using 2-photon excitation, Opt. Commun. Vol. 93(No. 5-6): 277–282.

12. Hillenbrand, R. & Keilmann, F. (2000). Complex optical constants on a subwavelength scale, Phys. Rev. Lett. Vol. 85(No. 14): 3029–3032.

13. Hillenbrand, R. & Keilmann, F. (2002). Material-specific mapping of metal/semiconductor/dielectric nanosystems at 10 nm resolution by backscattering near-field optical microscopy, Appl. Phys. Lett. Vol. 80(No. 1): 25–27

14. Hillenbrand, R., Stark, M. & Guckenberger, R. (2000). Higher-harmonics generation in tapping-mode atomic-force microscopy: Insights into the tip-sample interaction, Appl. Phys. Rev. Vol. 76(No. 23): 3478–3480.

15. Huang, H.-T. & Terry Jr., F. L. (2004). Spectroscopic ellipsometry and reflectometry from gratings (Scatterometry) for critical dimension measurement and in situ, real-time process monitoring, Thin Solid Films Vol. 455-456(No. 1-2): 828–836.

16. Jackson, J. D. (1975). Classical Electrodynamics, John Wiley, New York.

17. Keilmann, F. & Hillenbrand, R. (2004). Near-field microscopy by elastic light scattering from a tip, Phil. Trans. R. Soc. Lond. A Vol. 362(No. 1817): 787–805

18. Keller, D. (1991). Reconstruction of STM and AFM images distorted by finite-size tips, Surf. Sci. Vol. 253(No. 1-3): 353–364.

19. Klar, T. A., Engel, E. & Hell, S. W. (2001). Breaking Abbe's diffraction resolution limit in fluorescence microscopy with stimulated emission depletion beams of various shapes, Phys. Rev. E Vol. 64(No. 6): 066613.

20. Knoll, B. & Keilmann, F. (1999a). Mid-infrared scanning near-field

optical microscope resolves 30 nm, Journal of Microscopy Vol. 194(No. 2-3): 512–515.

21. Knoll, B. & Keilmann, F. (1999b). Near-field probing of vibrational absorption for chemical microscopy, Nature Vol. 399(No. 6732): 134–137.

22. Liu, H., Klonowski, M., Kneeburg, D., Dahlen, G., Osborn, M. & Bao, T. (2005). Advanced atomic force microscopy probes: Wear resistant designs, J. Vac. Sci. Technol. B Vol. 23(No. 6): 3090–3093

23. Mertz, J., Hipp, M., Mlynek, J. & Marti, O. (1994). Facts and artifacts in near-field microscopy, Appl. Phys. Lett. Vol. 81(No. 18): 2338–2340.

24. Novotny, L. & Hecht, B. (2006). Principles of nano-optics, Cambridge Univeristy Press, Cambridge. Ocelic, N. & Hillenbrand, R. (2004).

25. Subwavelength-scale tailoring of surface phonon polaritons by focused ion-beam implantation, Nature Materials Vol. 3(No. 9): 606–609.

26. Ocelic, N., Huber, A. & Hillenbrand, R. (2006). Pseudoheterodyne detection for background-free near-field spectroscopy, Appl. Phys. Lett Vol. 89(No. 10): 101124.

27. Porto, J. A., Carminati, R. & Greffet, J. J. (2000). Theory of electromagnetic field imaging and spectroscopy in scanning near-field optical microscopy, J. Appl. Phys. Vol. 88(No. 8): 4845–4850.

28. Thiault, J., Foucher, J., Tortai, J. H., Joubert, O., Landis, S., & Pauliac, S. (2005).

29. Line edge roughness characterization with a three-dimensional atomic force microscope: Transfer during gate patterning processes, J. Vac. Sci. Technol. B Vol. 23(No. 6): 3075–3079.

30. Xie, C., Mu, C., Cox, J. R. & Gerton, J. M. (2006). Tip-enhanced fluorescence microscopy of high-density samples, Appl. Phys. Lett. Vol. 89: 143117.

31. Zayats, A. & Richards, D. (2009). Nano-optics and near-field optical microscopy, Artech House, Norwood. Zenhausern, F., Martin, Y. & Wickramasinghe, H. K. (1995). Scanning interferometric apertureless microscopy: Optical imaging at 10 Angstrom resolution, Science Vol. 269(No. 5227): 1083–1085.

Chapter 2

FLUIDIC OPTICAL DEVICES BASED ON THERMAL LENS EFFECT

Hong Duc Doan and Kazuyoshi Fushinobu[1]

[1]Department of Mechanical and Control Engineering, Tokyo Institute of Technology, Meguro-ku, Tokyo, Japan

INTRODUCTION

Gordon et al. [1] reported that the beam shape of incident laser light expands after passing through a liquid medium. This phenomenon was termed "the thermal lens effect," and it has become a well-known photo-thermal phenomenon. Phenomenological, optical, and spectroscopic studies of the thermal lens effect have been carried out to describe nonlinear defocusing effect [2-8]. Recent progress in laser technology has revealed the various aspects of the thermal lens effect. Based on these efforts, other mechanisms, such as liquid density, electronic population, and molecular orientation, have been found to play important role as well as thermal lens effect. Recent studies term these effects as "the transient lens effect" [9,10]. The main advantage of using the transient lens effect in Photo-Thermal-Spectroscopy is that the sensitivity is 100 to 1000 greater than a traditional absorptiometry [11].

In this research, a new idea of applying the thermal lens effect in order to develop fluidic optical device is proposed. A schematic of the concept is shown in Fig. 1. A rectangular solid region shown inFig. 1a represents the liquid medium, which has a temperature field generated by a heater-heat sink system or laser-induced absorption. By controlling the temperature field as well as the refractive index distribution of the liquid medium, the refractive angle of each light ray passing through the liquid medium can be controlled in order to develop fluidic optical devices such as: an optical switching inFig. 1a to change the direction of the input laser beam, a laser beam shaper in Fig.

1b to transform a Gaussian beam to a flat-top beam and a fluidic divergent lens in Fig. 1c. Merits of these devices include flexibility of optical parameters, versatility and low cost.

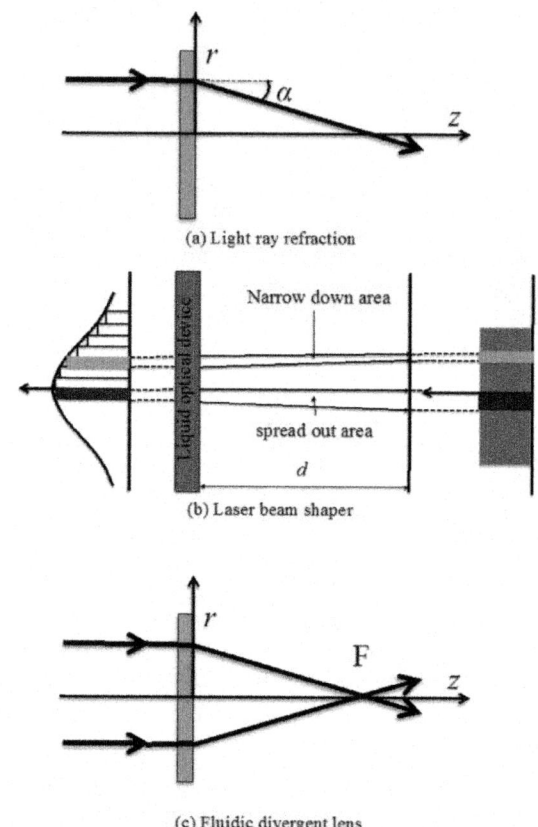

(a) Light ray refraction

(b) Laser beam shaper

(c) Fluidic divergent lens

Figure 1: Schematic of the concept of the fluidic optical devices.

FUNDAMENTAL LIGHT RAY TRANSMITTED IN ONE-DIMENSIONAL REFRACTIVE INDEX MEDIUM

In this section, as a first step to develop fluidic optical device, the refractive characteristics of a probe beam, which is transmitted in one-dimensional temperature distribution in a liquid medium is presented.

Theoretical Background

The light ray is modeled in the domain shown in Fig. 2 in order to calculate the refractive angle of the probe beam, which is transmitted in a one-dimensional temperature distribution in the liquid medium. The light ray direction transmitted in a medium having a refractive index dependent only on the y-axis, is described by the following form [12]

$$x = \int_0^y \frac{n_0 \sin\theta \cos\varphi}{\sqrt{n^2(y) - n_0^2 \sin^2\theta}} dy$$

(1)

$$z = \int_0^y \frac{n_0 \sin\theta \sin\varphi}{\sqrt{n^2(y) - n_0^2 \sin^2\theta}} dy$$

(2)

In which, the light ray passes through the medium at coordinate center, θ and φ are the incident angle with y and x-axis respectively, n_0 is the refractive index of the medium at the coordinate center.

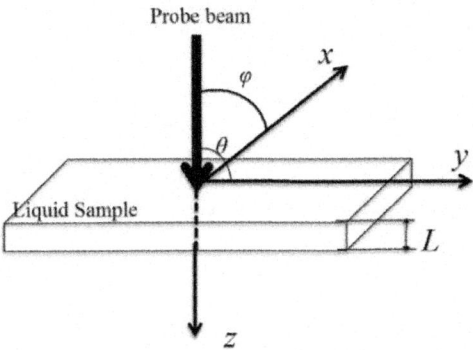

Figure 2: Schematic diagram of computational domain.

Figure 2 shows a schematic diagram of the model set up. The rectangular solid medium in the Figure represents the domain considered in the calculation which consists of ethylene glycol. In the medium, ethylene glycol has linear temperature distribution in only y-axis direction and the probe beam propagate along z-axis direction. Therefore, $\theta = \varphi = \pi/2$, and Eq. (1), (2) become:

$$X = 0$$

(3)

$$z = \int_0^y \frac{n_0}{\sqrt{n^2(y) - n_0^2}} dy$$

(4)

The temperature distribution of the liquid medium is modeled with a linear function of the variable yand temperature T at point y is calculated following:

$$T(y) = T_0 + \frac{dT}{dy} y$$

(5)

In which, T_0 is the temperature of the liquid medium at coordinate origin and dT/dy is constant.

Furthermore, between 0°C and 100°C refractive index of ethylene glycol is a linear function of temperature with refractive index change dn/dT = -2.6×10^{-4} 1/K [13]. Therefore, the relationship between refractive index and the variable y can be rewriten as follows:

$$y = \frac{n(y) - n_0}{dn / dy}$$

(6)

And

$$dy = dn \times \frac{dy}{dn} = dn \times \frac{dy}{dT} \times \frac{dT}{dn} = \frac{dn}{k}$$

(7)

Where,

$$k = -2.6 \times 10^{-4} \times \frac{dT}{dy}$$

(8)

By substituting Eq. (7) into Eq. (4) and solving the differential equation, we can obtain the relationship between y and z as:

$$y = \frac{n_0 [\exp(kz / n_0) - 1]^2}{2k \exp(kz / n_0)}$$

(9)

And refractive angle (RA) can be obtained as:

$$\alpha = \frac{dy}{dz} = \frac{1}{2} \exp(\frac{kz}{n_0}) - \frac{1}{2} \exp(-\frac{kz}{n_0})$$

(10)

Equation (10) shows the expression of the refractive angle as a function of the temperature gradient and the thickness of the liquid medium (optical path length). Figure 3 shows the relationship between refractive angle and temperature gradient at points where the thickness of sample, L, is 1.5, 3.0 and 4.5 mm respectively. As shown in Fig. 3 the relationship between the refractive angle and temperature gradient can be well approximated as linear at a small temperature gradient.

EXPERIMENTAL SET-UP

Figure 4 shows the experimental set-up to measure the refractive angle. Fluidic optical device in Fig. 4 is a pyrex vessel (internal size: 21×10×L [mm], thickness of the liquid medium, L, can be varied) filled with ethylene glycol. The vessel is held in an adiabatic material (Mica glass-ceramics (Photoveel®)) as shown in Fig. 5. The temperature on both sides of the vessel was controlled by a heater-heat sink system to create a one-dimensional temperature distribution in the liquid medium.

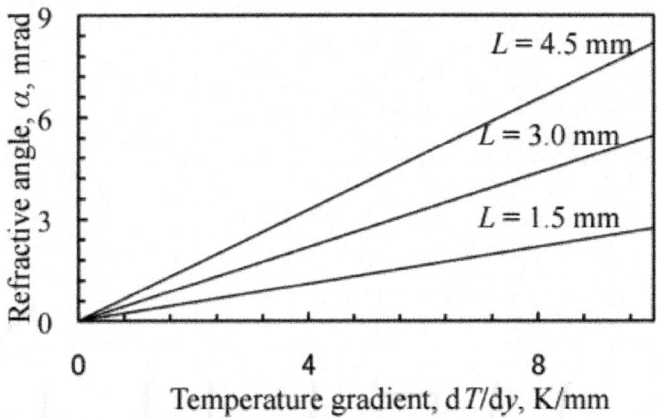

Figure 3: Relationship between the refractive angle and temperature gradient.

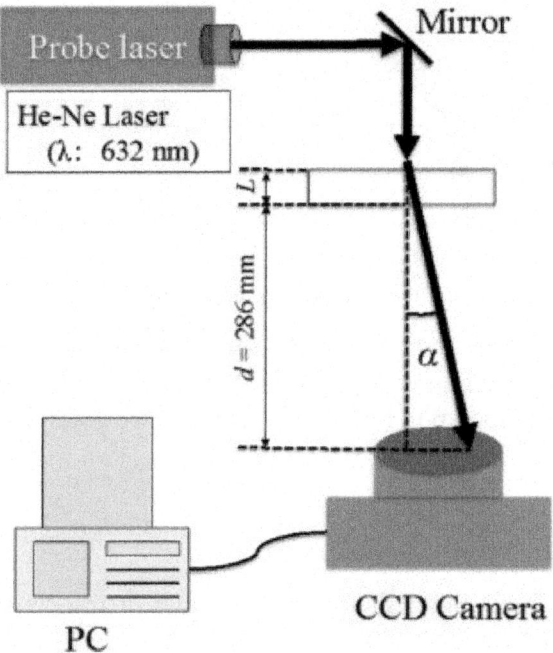

Figure 4:.Refractive angle measurement system.

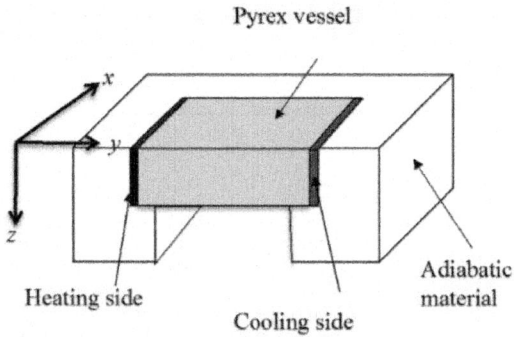

Figure 5:.Schematic diagram of sample.

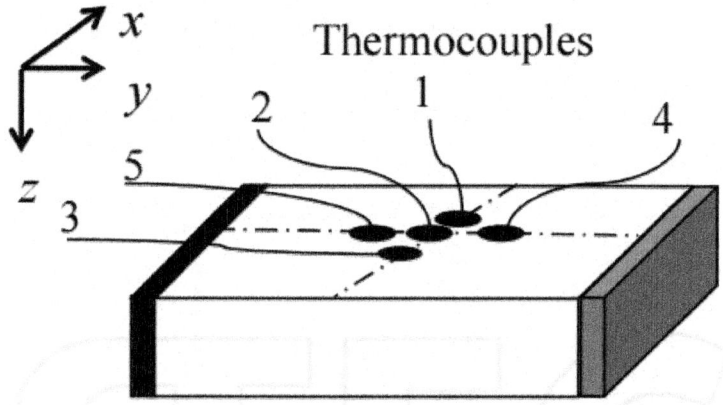

Figure 6: Temperature gradient measurement system.

The temperature distribution in the liquid medium is confirmed by measuring the temperatures at 5 points with 2 mm pitch in the vessel using 5 thermocouples as shown in Fig. 6. The temperature gradient in the experiment is given as:

$$\frac{dT}{dy} = \frac{T_4 - T_5}{\Delta y}$$

(11)

In which, Δy is the distance between two thermocouples to obtain the temperature gradient, $\Delta y = 4$ mm.

A CW laser ($P = 0.6$ mW, $\lambda = 632$ nm, $\Phi = 0.8$ mm, TEM00) is used as a probe beam. A CCD camera (OPHIR, BeamStar-FX 50) is used as a detector. Based on probe beam position, the refractive angle is estimated with:

$$\alpha = \frac{r}{d}$$

(12)

Where d is the distance from the sample to the detector of the camera = 286 mm; r is the beam position.

Results And Discussions

Figure 7(a) and (b) show the comparison of theoretical and experimental results at points where the thickness of sample, L, is 1.5 and 3.0 mm respectively. The temperature gradient in the theoretical results is based on the measurement as described above. As shown in this Figure, theoretical and experimental results agree well with each other. The experimental data includes the error corresponding to the difference between the actual temperature gradient at the laser incidence which is calculated by using Eq. (11). The discrepancy at higher dT/dy may correspond to the error where the measured temperature gives higher dT/dy and therefore higher a prediction by using Eq. (10). This discrepancy should increase with increasing sample thickness and the temperature gradient as a consequence of the effect of natural convection and the temperature gradient in the z-axis [14].

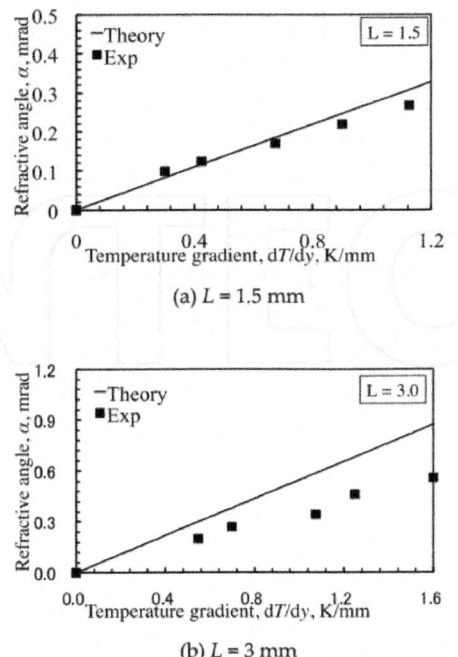

(a) L = 1.5 mm

(b) L = 3 mm

Figure 7: The comparison of theoretical and experimental results

FLUIDIC LASER BEAM SHAPER

Flat-top laser are well known to present significant advantages for laser technology, such as holographic recording system, Z-scan measurement, laser heat treatment and surface annealing in microelectronics and various nonlinear optical processes [15-19]. For CW beams, several approaches to spatially shape Gaussian beams have been developed, such as the use of aspheric lenses, implement beam shaping or the use of diffractive optical devices [20]. However, these methods have some disadvantages: a refractive beam shaping system lead to large aberration [21] and implemental beam shaping has low energy efficiency and lacks of flexibility [22]; and the use of refractive optical devices requires complex configuration design and high cost [23]. In practice, a low-cost and flexible method to convert a Gaussian beam into a flat-top beam is required. In this section, a novel method to convert a Gaussian beam into a flat-top beam is discussed. The concept is based on the control of the pump power and propagation distance of the probe beam in the thermal lens system.

Principle Of Thermal Lens Effect

The principle of the transient lens effect is schematically illustrated in Fig.8. A CW diode pumped blue laser is used as pump-beam (BCL-473-030, $\lambda = 473$ nm, $\Phi = 0.8$ mm, TEM_{00}) with maximum output power of 30mW. The laser beam intensity was adjusted by using ND-filter. A CW infrared DPSS laser is used as probe-beam (MIL, $\lambda = 1064$ nm, $\Phi = 3.0$ mm, TEM_{00}) with maximum output power of 10 mW. A CCD camera is used as a detector to measure the intensity distribution of the laser beam. A cuvette, which is a three-layer structure with a sheet copper is sandwiched between 2 pieces of fused silica. The height of the fused silica is 1 mm. The sheet copper has doughnut shape. The liquid that is contained inside the doughnut hole has the same height with the sheet copper. By varying the thickness of the sheet copper, the liquid height can be changed. The ethanol solution dissolved dye termed as Sunset-yellow is filled in the cuvette. The chemical formula of the Sunset-yellow is shown in Ref. 24.

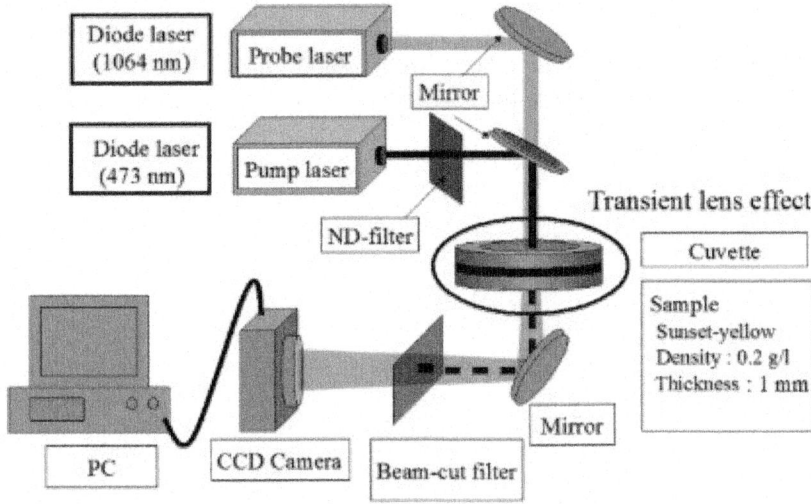

Figure 8: Schematic diagram of a dual-beam thermal lens system.

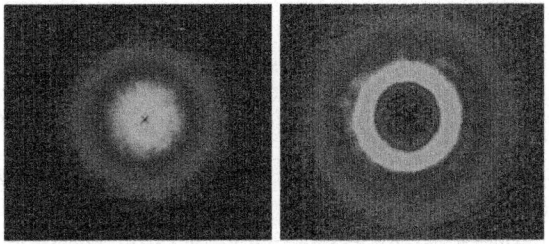

Figure 9: Experiment result: the difference of intensity profile of the probe beam after passing through the thermal lens (left side) and a quart divergent lens (right side)

In this experiment, the absorbance of the pump-beam is 2.776 and that of the probe-beam is negligible small. Figure 9 shows the laser beam profile of the probe beam after propagating through a divergent lens and a thermal lens. It is clear that, the probe beam change its profile from Gaussian to doughnut beam with a hollow center is created.

Theoretical analysis of laser beam profile change in thermal lens effect is done with a model that includes continuity equation, Navier-Stokes equation, energy conservation equation and Helmholtz equation in 2D cylindrical symmetry coordinate. It is assumed that the change of refractive index is caused only by the temperature change of the liquid medium and the thermal coefficient of the refractive index, dn/dT. The concentration is supposed to be constant over the range of the temperature rise induced by the pump beam.

When the liquid medium is irradiated, temperature distribution perpendicular to the optical axis is formed due to intensity distribution of laser beam and heat transport. To consider the natural convection effect, the temperature distribution of liquid sample in steady state is calculated numerically following these governing equations [24]:

$$\frac{1}{r}\frac{\partial}{\partial r}(rv_r) + v_z\frac{\partial v_z}{\partial z} = 0 \tag{13}$$

$$v_r\frac{\partial v_r}{\partial r} + v_z\frac{\partial v_r}{\partial z} = -\frac{1}{\rho}\frac{\partial p}{\partial r} + v\left(\frac{\partial}{\partial r}\left(\frac{1}{r}\frac{\partial}{\partial r}(rv_r)\right) + \frac{\partial^2 v_r}{\partial z^2}\right) \tag{a}$$

$$v_r\frac{\partial v_z}{\partial r} + v_z\frac{\partial v_z}{\partial z} = -\frac{1}{\rho}\frac{\partial p}{\partial z} + v\left(\frac{1}{r}\frac{\partial}{\partial r}\left(r\frac{\partial v_z}{\partial r}\right) + \frac{\partial^2 v_z}{\partial z^2}\right) + g\beta(T - T_0) \tag{b}$$

$$\tag{14}$$

$$v_r\frac{\partial T}{\partial r} + v_z\frac{\partial T}{\partial z} = a\left(\frac{1}{r}\frac{\partial}{\partial r}\left(r\frac{\partial T}{\partial r}\right) + \frac{\partial^2 T}{\partial z^2}\right) + S \tag{15}$$

$$S = \frac{\alpha e^{-\alpha z}I_0(r)}{\rho C_p} \tag{16}$$

Here, $I_0(r)$ is the intensity distribution of the pump laser. The spot sizes of the laser beams are assumed to be constant through the interaction volume within the liquid medium.

The temperature distribution is calculated numerically based on the finite difference method. The 1st order upwind scheme and a 2nd order center differencing are applied to discretize the advection term and the diffusion term respectively. The thermal properties of liquid medium can be found in Ref. 24.

To model the propagation of laser through an inhomogeneous medium, the wave equation which includes an absorption term and an inhomogeneous refractive index term is applied [24]:

$$-\frac{\partial^2 E}{\partial z^2} + 2ik_0n_0\frac{\partial E}{\partial z} = \frac{1}{r}\frac{\partial}{\partial r}\left(\frac{\partial(rE)}{\partial r}\right) + k_0^2\left(n^2 - n_0^2\right)E - \frac{1}{2}ik_0n_0\alpha E \tag{17}$$

Here, E is the envelope of the oscillating electric field, z is the axis of propagation, r is transverse coordinates, k_0 is the free space wave number and α is the absorption coefficient. The variable n is the refractive index profile depending on medium temperature following:

$$n(T) = n_0 + \frac{dn}{dT}(T - T_0) \tag{18}$$

Here, $n_0 = 1.359$ is the refractive index of the liquid medium at reference temperature $T_0 = 298.15$ K, dn/dT is the temperature coefficient of the refractive index. The propagation of laser is calculated based on Pade method. The optical properties parameter can be found in Ref. 24.

Influences Of The Pump Power And The Propagation Distance On The Change Of Probe Beam Profile

Influences of the pump power and the propagation distance to the probe beam profile were investigated numerically using the calculation parameters in Table. 1. In this calculation, both of the pump beam and the probe beam are written as follows.

$$E = E_0 \exp\left(\frac{-r^2}{r_0^2}\right) \exp\left(\frac{-ik_0 n_0 r^2}{2R}\right)$$

(19)

$$E_0 = \sqrt{\frac{2P}{\pi r_0^2}}$$

(20)

Here P is the power of laser, R is the radius of curvature of the wave front, r and r0 are distance from laser axis and beam radius respectively.

Table 1: Calculation conditions

Parameter	(a)	(b)
Pump power, mW	3	0 ~ 7
Pump beam diameter, mm	0.8	0.8
Probe power, mW	10	10
Probe beam diameter, mm	0.8	0.8
Absorption coefficient, cm^{-1}	2.0	2.0
Distance from experimental section to CCD camera, mm	0 ~ 500	200
Phase front curvature radius, R, mm	∞	∞

Effects of the pump power and the propagation distance to the probe beam profile are shown in Fig. 10(a) and (b) respectively. The vertical axis and horizontal axis show intensity and distance from laser axis respectively. Plots of 'P = 0 mW' and 'd = 0 mm' represent intensity distribution of the probe beam without thermal lens effect. As shown in Fig. 10(a), the further the propagation distance, the lower intensity at the probe beam center, and higher intensity at the wing. With increasing of the propagation distance, the laser beam profile changes from Gaussian to flat-top and the doughnut beam profile respectively. The profile of the probe beam changes with the same tendency as the increasing of pump power as shown in Fig. 10(b). In particular, when the pump power is 3 mW and propagation distance is 200 mm the probe beam is converted to the flat-top profile approximately. Therefore, by controlling the pump power and the propagation distance the Gaussian beam can be converted into the flat-top beam.

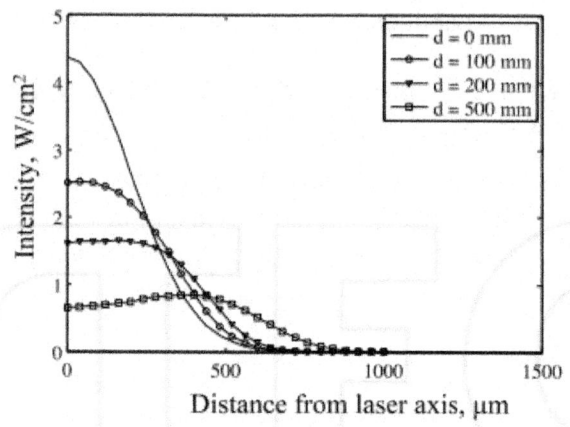

(a) Influence of the propagation distance

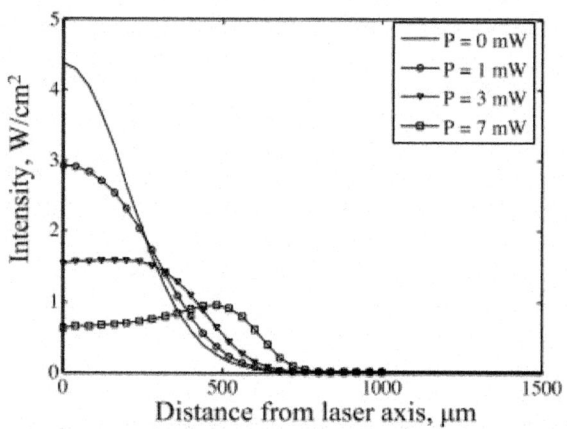

(b) Influence of the pump power

Figure 10: Influence of the propagation distance and the pump power to the probe beam profile

Experimental Set-Up To Shape Spatial Profile

In order to confirm the role of the fluidic laser beam shaper, a single-beam experiment is set up as shown in Fig. 11. A CW diode blue laser is used as pump and probe-beam (P = 10 mW, λ = 488 nm, Φ = 0.69 mm, TEM00). In this experiment, the height of the liquid medium is 0.5 mm, the dye concentration is 0.1 g/l and the absorption coefficient is 2.92 cm^{-1} (measured value) respectively. The propagation distance to obtain the flat-top beam profile is measured by changing the distance from the cuvette to the CCD camera. At the propagation distance of 150 mm, the flat-top beam is confirmed as shown in Fig. 12(a).

Figure 12(b) shows the beam profile change from the Gaussian to the flat-top beam. The vertical and horizontal axes show the intensity and distance from the laser axis respectively. The o-line shows the profile of the Gaussian input beam by fitting the laser beam profile measured at the surface of the cuvette. The strange-line shows the profile of the flat-top beam calculated by beam propagation method. The solid-line shows the profile of the flat-top beam measured by CCD camera at propagation distance of 150 mm from the cuvette. Both experimental and calculated results agree well with each other.

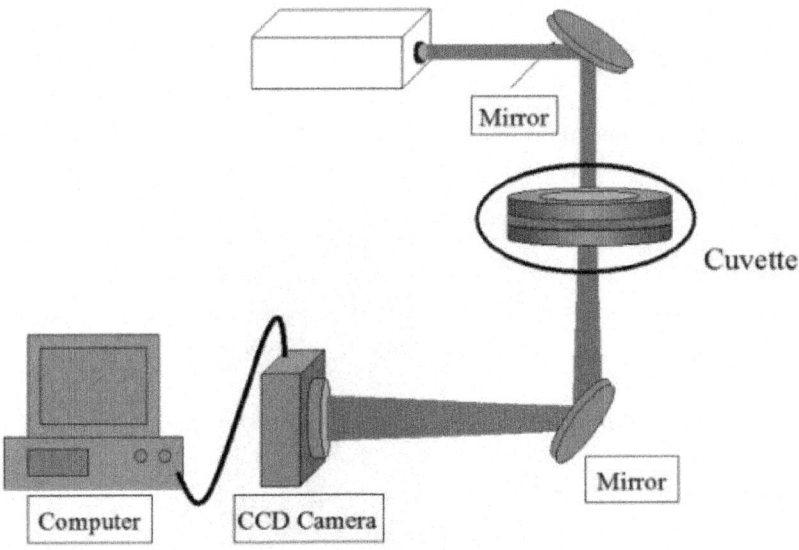

Figure 11: Experimental set up for a single-beam thermal lens system to transfer a Gaussian beam to flat-top beam

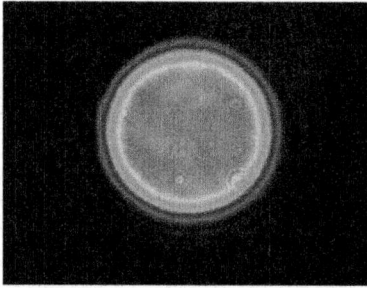

(a) Flat-top beam profile measured by CCD camera

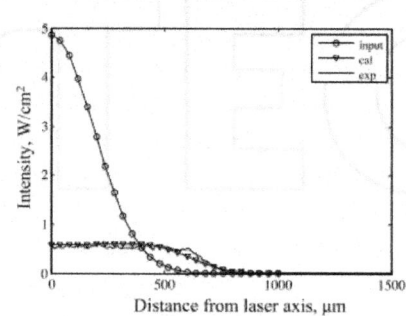

(b) Beam profile change from Gaussian to flat-top

Figure 12: Experimental results.

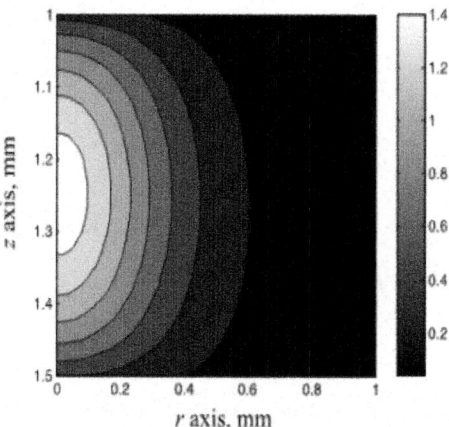

Figure 13: Temperature distribution inside the liquid medium [K]. The calibration shows the difference between temperature inside the liquid medium with the ambient temperature.

In order to explain in more detail about the mechanism of this fluidic beam shaper, the temperature distribution of liquid medium is calculated. As shown in Fig. 13, local heating near the beam axis produces a radially dependent temperature variation, which changes the liquid refractive index in which the lower refractive index is in the region near to the beam center. As a consequence, the radius of curvature of the wave front at the region near the beam center is shorter than one at the beam wing. Therefore the sample liquid locally acts as a micro divergent lens with shorter focal length at beam center. As shown in Fig. 1b, the beam center that passes through shorter focal length is spread out more rapidly than the beam wing. As the probe beam propagates to increasing distance, the intensity in the center region drops rapidly than one in the wing region. At a certain value of propagation distance, the Gaussian beam can be converted into the flat-top beam.

It is noted that, in the case of single-beam shaper, one part of laser beam energy (about 15% in this experiment) is converted into thermal energy in order to change temperature distribution or in other words to change refractive index distribution in the liquid medium. Therefore, in the case of single-beam shaper, the beam shaper has another role, which is as an attenuator. This laser beam shaper/attenuator can be applied in practical laser drilling technology. In the case of applying on only laser beam shaper, the double-beam system is recommended. In this case, it is needed to select dye whose absorbance of the probe-beam is negligible small.

Relationship Between Pump Power And Distance To Shape Spatial Profile

As shown in previous section, the flat-top beam can be obtained only at a fixed distance. In order to control this distance, the influence of pump power is investigated theoretically and experimentally. The calculation parameters are shown in Table. 2. The pump power is changed from 1 to 8 mW. The distance to obtain the flat-top beam is obtained numerically. The relationship between the pump power and the distance to shape spatial profile is shown in Fig. 14(a). The horizontal and vertical axes show pump power and distance to obtain the flat-top beam respectively. As shown in Fig. 14(a), the distance to obtain a flat-top beam is in inverse proportion to the pump power.

In order to validate the numerical prediction, a single beam experiment was carried out. The pump power is changed from 1 to 6 mW and the distance to obtain the flat-top beam was measured. The experimental result shown in Fig. 14(b), shows excellent agreement with calculation prediction. The relationship between pump power and distance to obtain the flat-top beam can be explained by the interaction between energy absorption of liquid medium with the focal length of local micro lens. As the pump power increase, the absorption energy increases. As a consequence, the rate of decreasing ofR is enhanced. This can be thought as the reason why the distance to obtain a flat-top beam decreases. In other words, the distance to obtain the flat-top beam profile also decreases with the increasing of absorption coefficient. Therefore, by changing the absorption coefficient or the pump power, the distance to obtain a flat-top beam can be controlled.

Table 2: Calculation conditions

Pump power, mW	1 ~ 8
Pump beam diameter, mm	0.8
Probe power, mW	10
Probe beam diameter, mm	0.8
Absorption coefficient, c^{m-1}	2.0
Phase front curvature radius, R, mm	320

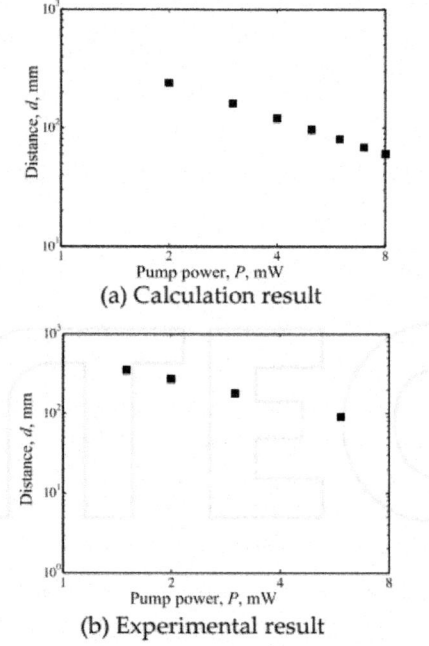

(a) Calculation result

(b) Experimental result

Figure 14: Relationship between the pump power and the distance to obtain the flat-top beam profile. The horizontal and vertical axes show the pump power and the distance to obtain the flat-top beam profile respectively.

TUNABLE FLUIDIC LENS

Fluidic lenses are well known to present significant advantages for wide range of applications from mobile phone to laboratory on a chip. Fluidic lenses have a number of apparent advantages such as tunable refractive index and reconfigurable geometry. Several approaches to design the liquid lens have been developed based on the microfluidic techniques to modify the liquid lens shape by using: out-of-plane micro-optofluidic [25-26], in-plane micro-optofluidic [27-28], electron wetting [29], dielectrophoresis [30] and hydrodynamic force [31]. Other approach bases on turning the refractive index of the liquid by different means such as pressure control, optical control, magnetic control, thermo-optic control, and electro-optic control.

Principle Of Fluidic Lens

When the liquid medium is irradiated, local heating near the beam axis produces a radially dependent temperature variation, which changes the liquid refractive index in which the lower refractive index is in the region near to the beam center. As a consequence, the radius of curvature of the wave front at the region near the beam center is shorter than one at the beam wing. The liquid medium behaviors as a convergence GRIN-L with focal length depends on the radial position of the incident ray relative to the optical axis of the cuvette. The ray equation that is calculated numerically to obtain the path of an incident beam, which is given by:

$$\frac{d}{ds}\left(n\frac{dR}{ds}\right) = \text{grad}(n)$$

(21)

Where, ds and R are the differential element of the path length and the positional vector of the ray respectively. The variable n is the refractive index of the liquid sample. The variable n is the refractive index profile depending on medium temperature following equation (13-16, 18).

Influences Of The Pump Beam Profile

Influences of the pump beam profile to the focal length of the GRIN-L were investigated numerically with the calculation conditions in Table. 3. The intensity profile of the pump is applied with the Gaussian beam and the quasi-flat-top beam (a super-Gaussian distribution of order k) using Eq. 22 andEq. 23 respectively.

$$I_{Gaussian} = \frac{2P}{\pi r_0^2}\exp\left(\frac{-2r^2}{r_0^2}\right)$$

(22)

$$I_{\text{Flat-top}} = \frac{Pk2^{2/k}}{2\pi r_0^2 \Gamma(2/k)} \exp\left(\frac{-2r^k}{r_0^k}\right)$$

(23)

Here Γ is the Gamma function, r and r0 are distance from laser axis and beam radius, respectively. In this calculation, quasi-flat-top beam is the 10 order of the super-Gaussian distribution, two types of the pump intensity profile are shown in Fig.15.

Table 3: Calculation conditions

Pump power, mW	10
Pump beam diameter, mm	1.5
Absorption coefficient, c^{m-1}	2.0

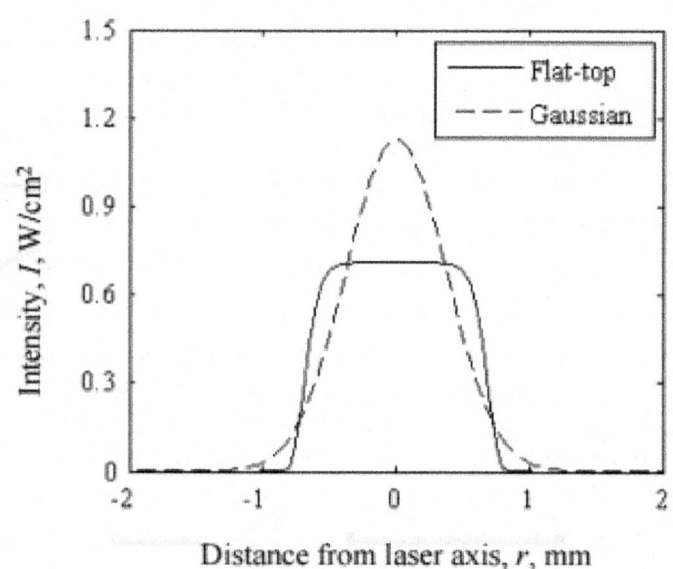

Figure 15: Two types of the pump beam profile.

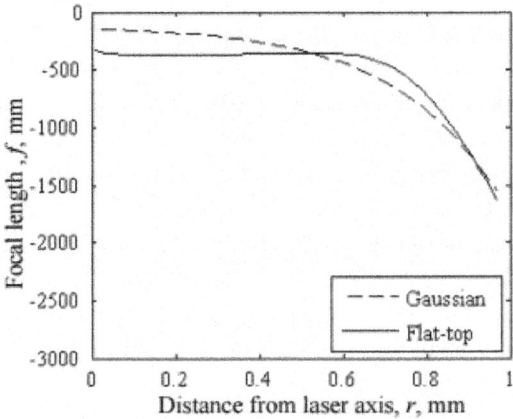

Figure 16: Effect of the pump beam profile to the focal length of the GRIN-L lens.

The effect of the pump beam profile to the focal length of the GRIN-L lens is shown in Fig. 16. The vertical and horizontal axes show focal length and distance from laser axis respectively. The solid and dashed lines represent the plot of the focal length again the radial position of the incident ray relative to the optical axis of the cuvette in the case of Gaussian pump beam and quasi flat-top pump beam respectively. As shown in Fig. 16, for the Gaussian pump beam the focal length of the GRIN-L increases sharply with increasing of the distance from laser axis, which means larger spherical aberration. It means that, the beam center which passes through shorter focal length is spread out more rapidly than the beam wing. As a consequence, the further the propagation distance of the probe beam, the laser beam profile changes from Gaussian to the doughnut beam profile [24], which should cause some undesirable results in laser processing [32]. In contrast, with the quasi flat-top pump beam, the focal length of the GRIN-L varies lightly with increasing of the distance from laser axis smaller than beam waist of the flat-top pump beam. The area smaller than the beam waist of the flat-top pump beam acts as a divergent lens with small spherical aberration. Therefore, for the purpose of designing the GRIN-L lens the uniform pump beam shows the advance in reducing the spherical aberration.

Experimental Set-Up

In order to confirm the qualities of the GRIN-L, an experiment with the quasi flat-top pump beam is carried out as shown in Fig. 17. A CW diode blue laser is used as pump laser (P = 10 mW, = 488 nm, = 0.69 mm, TEM00). In cuvette 1, the height of liquid is 0.5 mm, and the absorption coefficient is 2.92 cm^{-1} (at wavelength of 488 nm). In the cuvette 2, the height of liquid is 1 mm, and the absorption coefficient is 55 cm^{-1} (at wavelength of 488 nm). A CW He-Ne laser is used as probe laser (P = 0.6 mW, λ = 632 nm, Φ = 0.8 mm, TEM00). It is noted that, the absorption of ethanol solution can be ignored at the wavelength of the probe laser. First, the pump beam passes through cuvette1, then the beam profile of pump beam was converted from Gaussian to flat-top during its transmission to cuvette 2 as shown in Fig. 18. Then, the probe laser was adjusted to overlap with pump laser. After propagating through the sample the probe laser is directed towards the CCD camera and the pump laser is blocked using filters located at the detection plane. The distance between cuvette 2 and the CCD camera is varied, and the 1/e^2 diameter of probe laser is measured.

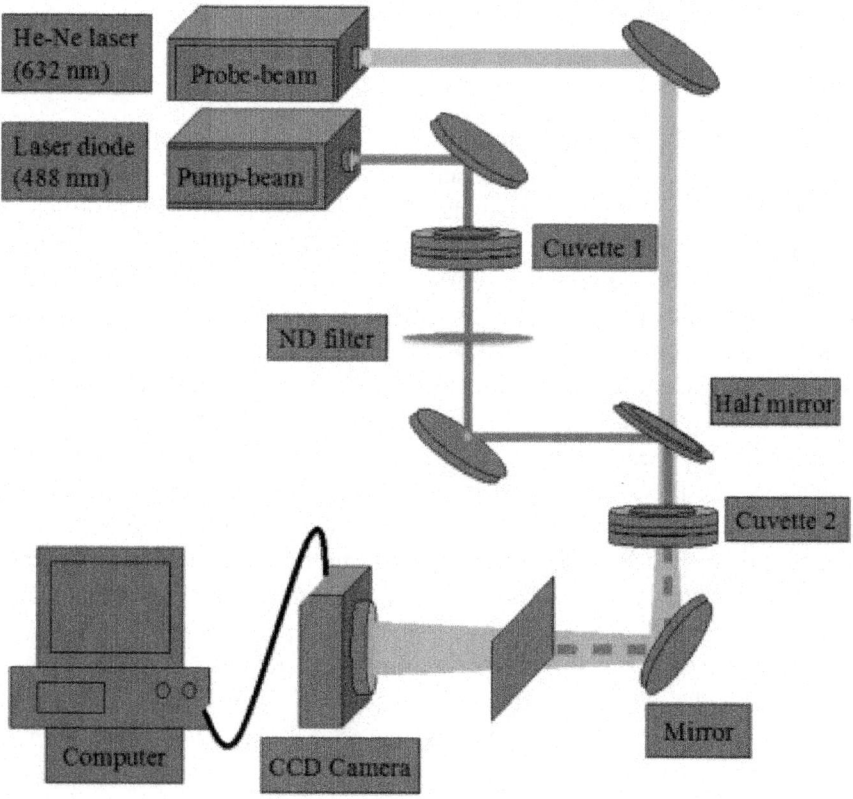

Figure 17: Experimental set-up for the fluidic divergent lens.

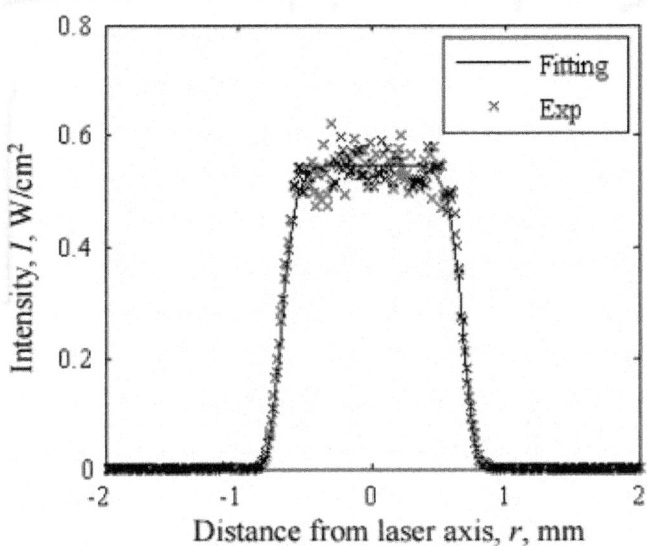

Figure 18: The intensity profile of the pump beam during its transmission to cuvette 2. Dotted and solid lines show the measured result and fitting by super-Gaussian distribution respectively.

Figure 19(a) shows the change along the propagation direction in the beam profile. The vertical and horizontal axes show the intensity and distance from the laser axis respectively. By using the quasi flat-top pump beam, the beam profile of probe laser can remain in Gaussian distribution during its propagation. Figure 19(b) shows the plot of probe beam waist again propagation distance. As shown in Fig. 19(b), the beam waist of probe laser varies linearly with propagation distance. In other words, cuvette 2 acts as a divergence lens with focal length of $f = -424$ mm (this value has been calculated by considering the divergence angle of probe laser $= 1.2$ mrad).

Next, the pump power is changed from $P_0 = 7.7$ mW to $P_0/2$, $P_0/3$ and $P_0/4$ respectively. Figure 20 shows the plots of focal length against the pump power. Square and circle plots show the calculation and experimental result, respectively. As shown in Fig. 20, the focal length increases with increasing of the pump power. This means that, by adjusting the pump power, the focal length can be controlled.

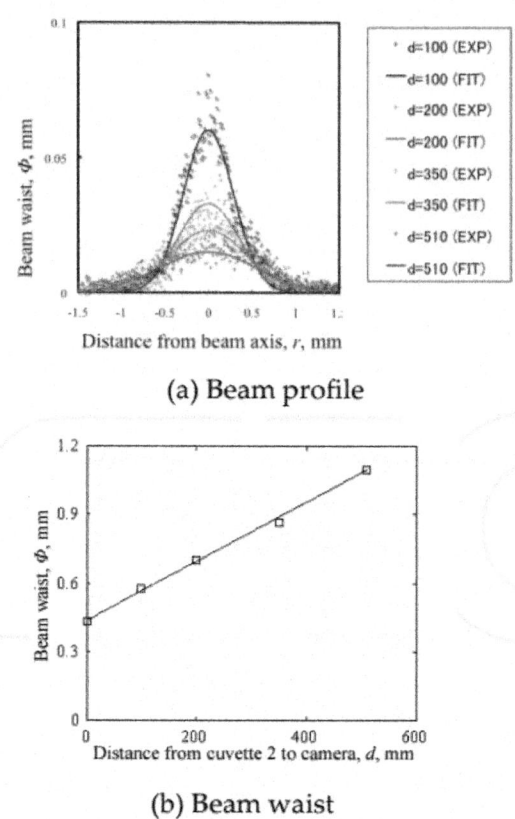

(a) Beam profile

(b) Beam waist

Figure 19: Probe beam changes along the propagation direction.

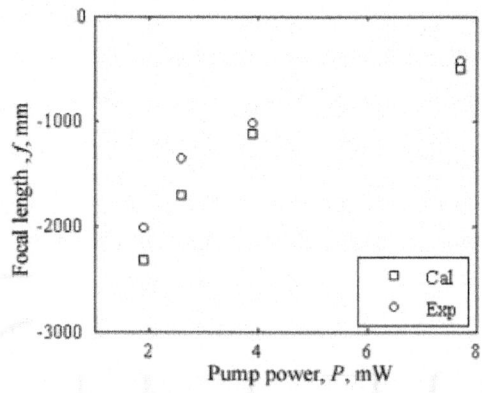

Figure 20: Relationship between the pump power and focal length.

CONCLUSION

In this research, a novel idea of fluidic optical devices which includes laser beam shaper and fluidic divergent lens are demonstrated. The fluidic optical devices are based on controlling some parameters in the thermal lens system. The interaction among the intensity distribution, power of the pump beam, the absorption coefficient, the propagation distance and the intensity profile of the probe beam have been investigated experimentally and theoretically. It is found that

By controlling the pump power and the absorption coefficient, the input Gaussian beam can be converted into a flat-top beam profile. The distance to get the flat-top beam profile can be controlled easily by adjusting the pump power and the absorption coefficient. In actual applications, single-beam shaper has another role, which is as an attenuator. This laser beam shaper/attenuator can be applied in practical laser drilling technology. In the case of applying on only laser beam shaper, the double-beam system is recommended. In this case, it is needed to select a dye whose absorbance of the probe-beam is negligible small.

The uniform pump beam shows the advance in reducing the spherical aberration. And by adjusting the pump power, the focal length can be controlled

With some merits such as flexiblility, versatility and low cost, these fluidic optical devices will be promising tools in many fields of laser application.

ACKNOWLEDGEMENT

Part of this work has been supported by the Grant-in-Aid for JSPS Fellows and Grant-in-Aid for Scientific Research of MEXT/JSPS. The authors also would like to acknowledge Mr. Akamine Yoshihiko.

REFERENCES

1.   J. P. Gordon, R. C. C. Leite, R. S. Moore, S. P. S. Porto, J. R. Whinnery, . Long-Transient, in. effects, with. lasers, liquid. inserted, J. samples, Appl. Phys. 36 (196538

2.   S. A. Akhmanov, D. P. Krindach, A. V. Migulin, A. P. Sukhorukov, R. V. Khokhlov, self-actions. Thermal, laser. of, I. E. E. E. J. beams, Quantum Electronics QE-419681968568

3.   P. M. Livingston, induced. Thermally, of. a. modifications, power. C. W. high, beam. laser, Appl, Opt. 1019711971426436

4.   J. F. Power, mode. Pulsed, lens. thermal, detection. effect, the. in, field. near, thermally. via, probe. induced, spatial. beam, modulation. a. phase,

Appl. theory, Opt. 291990199052

5. P. P. Banerjee, R. M. Misra, M. Maghraoui, Theoretical, studies. experimental, propagation. of, beams. of, a. through, sample. finite, a. of, nonlinear. cubically, J. material, Opt. Soc. Am. B 81991199110721080

6. J. M. Hickmann, A. S. L. Gomes, C. B. de Araújo, of. Observation, cross-phase. spatial, effects. modulation, a. in, nonlinear. self-defocusing, Phys. medium, Rev. Lett. 681992199235473550

7. P. Govind, Transverse. Agrawal, instability. modulation, copropagating. of, beams. optical, nonlinear. in, media. J. Kerr, Opt. Soc. Am. B 71990199010721078

8. C. J. Rosenberg, et al.of. Analysis, dynamics. the, high. of, Gaussian. intensity, beams. laser, nonlinear. in, Kerr. de-focusing, Optics. media, 2. Communications, 2007

9. M. Sakakura, M. Terazima, of. Oscillation, refractive. the, at. index, focal. the, of. a. region, laser. femtosecond, inside. a. pulse, Opt. glass, lett. 29 (13) (2004

10. M. Sakakura, M. Terazima, observation. Real-time, photothermal. of, after. effect, of. photo-irradiation, laser. femtosecond, inside. a. pulse, J. glass, Phys. France 12520052005355360

11. M. Terazima, N. Hirota, S. E. Braslavsky, A. Mandelis, S. E. Bialkowski, G. J. Diebold, R. J. D. Miller, D. Fournier, R. A. Palmer, A. Tam, terminology. Quantities, in. symbols, photothermal, spectroscopies. . I. U. P. A. C. related, Recommendations, 2004Pure Appl.Chem., 76, 1083

12. Kudou Uehara1990Basic Optics (Kougaku Kiso) Gendaikougaku, Tokyo, Japan, 4547Japanese)

13. S. K. Y. Tang, B. T. Mayers, D. V. Vezenov, G. M. Whitesides, waveguiding. Optical, thermal. using, across. gradients, liquids. homogeneous, microfluidic. in, Appl. channels, Phys. Lett. 88, 06112, 2006

14. H. D. Doan, K. Fushinobu, K. Okazaki, on. Investigation, interaction. the, light. among, material, field. temperature, the. in, lens. transient, transmission. effect, in. . D. characteristics, field. temperature, I. Proc, ITherm 2010127

15. J. Yang, Y. Wang, X. Zhang, C. Li, X. Jin, M. Shui, Y. Song, of. Characterization, transient. the, effect. thermal-lens, flat-top. using, Z-scan. J. beam, Phys. B: At. Mol. Opt. Phys. 42 (2009pp)

16. K. Ebata, K. Fuse, T. Hirai, K. Kurisu, laser. Advanced, for. optics, material. laser, Proc. S. P. I. processing, SPIE, 50632003

17. E. B. S. Govil, J. P. Longtin, A. Gouldstone, M. D. Frame, visible. Uniform-intensity, source. light, in. for, imaging. situ, of. Journal, Optics. Biomedical, 14(, 2009

18. F. M. Dickey, S. C. Holswade, D. L. , Shealy 'Laser Beam Shaping Applications', Taylor & Francis, 2006

19. M. T. Eismann, A. M. Tai, J. N. Cederquist, design. Iterative, a. of, beam. holographic, Appl. former, Opt. 281998199826411650

20. F. M. Dickey, S. C. Holswade, beam. Laser, Theory. shaping, Marcel. Techniques, New. Dekker, York, 2000

21. P. Scott, optics. Reflective, irradiance. for, of. redistribution, beam. laser, Appl. design, Opt. 20 (9) (1981

22. S. Zhang, Q. Zhang, G. Lupke, beam. Spatial, of. shaping, laser. ultrashort, theory. pulse, Appl. experiment, Opt.442005200558185823

23. B. Mercier, J. P. Rousseau, A. Jullien, L. Antonucci, beam. Nonlinear, for. shaper, laser. femtosecond, from. pulses, to. Gaussian, profile. flat-top, Communications. 2. Optics, 2010

24. H. D. Doan, Y. Akamine, K. Fushinobu, laser. Fluidic, shaper. beam, using. by, lens. thermal, Int. effect, J. Heat Mass Transfer (2012

25. S. H. Ahn, Y. K. Kim, of. Proposal, eye's. human, lens-like. crystalline, focusing. variable, Sens. lens, Actuators (1999A 78, 48

26. D. Y. Zhang, V. Lien, Y. Berdichevsky, J. H. Choi, Y. H. Lo, adaptive. Fluidic, with. lens, focal. high, tenability. length, Appl, Phys. Lett. 82 (2003

27. S. K. Hsiung, C. H. Lee, G. B. Lee, electrophoresis. Microcapillary, utilizing. chips, micro-lens. controllable, structures, optical. buried, for. fibers, optical. on-line, Electrophoresis. detection, 2008

28. V. Lien, Y. Berdichevsky, Y. H. Lo, surfaces. Microspherical, predefined. with, lengths. focal, using. fabricated, capillaries. microfluidic, Appl, Phys. Lett. (2003

29. C. B. Gorman, H. A. Biebuyck, G. M. Whitesides, of. Control, Shape. the, Liquid. of, on. a. Lenses, Gold. Modified, Using. Surface, Applied. an, Potential. Electrical, a. across, Monolayer. Self-Assembled, Langmuir, 1995

30. C. C. Cheng, C. A. Chang, H. A. Yeh, focus. Variable, liquid. dielectric, lens. droplet, Opt, Express (2006

31. S. K. Y. Tang, C. A. Stan, G. M. Whitesides, reconfigurable. Dynamically, liquid-cladding. liquid-core, in. a. lens, channel. microfluidic, Chip. Lab, 2008

32. D. H. Doan, Y. Yin, N. Iwatani, K. Fushinobu, processing. Laser, using. by, laser. fluidic, shaper. beam, Proc, National Heat Transfer Symposium 2012Inpress

Chapter 3

NOVEL OPTICAL DEVICE MATERIALS – MOLECULAR-LEVEL HYBRIDIZATION

Kyung M. Choi[1]

[1]University of California at Irvine, USA

INTRODUCTION

Silicate glass has been widely used for optical device materials due to its excellent optical transparency. To satisfy our multiple demands in advanced optical device materials, organic/inorganic hybrid composites have been widely prepared by a bulk mixing technique, which is physically mixing multiple components at the bulk scales.

However, conventional glassy materials have shown limitations to modify their physicochemical properties by inserting desired components into glassy hosts. In addition, a significant phase separation occurs during a mixing process of multiple components, especially immiscible phases.

To overcome those limitations, there are growing interests in doping organic components or semiconductor particles into glassy hosts without any phase separation to combine beneficial properties at the molecular scales, and thus to bring desired properties (Figure 1). [1]

Hybrid materials lie at the interface of the organic and inorganic material regimes, where versatility in molecular tailoring approach offers novel molecular modifications in design of new chemical structures. Hybrid materials can also range, depending on the method of formation and domain size, from physical mixtures of inorganic oxides and organics (blends, composites) to nanocomposites and molecular composites that utilize formal chemical linkages between the organic and inorganic domains on the molecular scale.

Hybrid materials are ranged from the bulk-scales to molecular scales as shown in Figure 2 to mix up multiple components. [1-10]

Usually, hybrid materials mixed at the bulk scales retain the original properties of the individual organic and inorganic components. In other words, their final properties are significantly influenced by the characteristics and their domain sizes of individual components after the mixing process at the bulk scales.

Organic-Inorganic Hybrid Materials

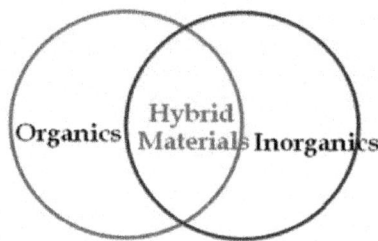

Combination of beneficial properties
At the molecular-level

Figure 1: Organic/inorganic hybrid materials. [1].

In addition, a phase separation problem also limits to achieve a uniform mixing of multiple components during the bulk-mixing process.

Hybrid Materials

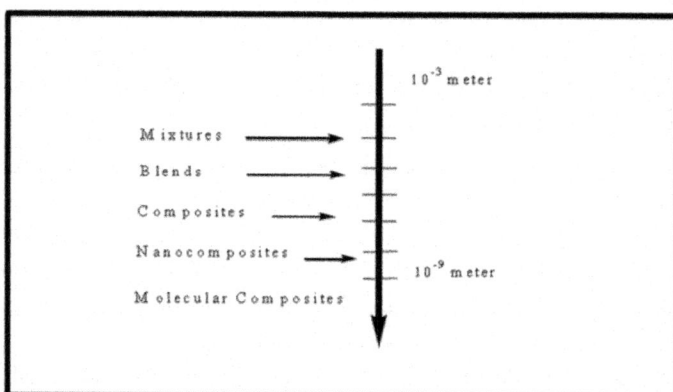

Figure 2: Relative size scale of mixing domains for different types of hybrid materials. [1-10].

To overcome those limitations of the bulk-scale mixing technique, molecular-level hybridization technique has been actively investigated. [1-10] This technique results in molecular-level composites, which are their domain sizes in the nanometer scale often create new properties, which would not be expected from those individuals by the loss of individuals' identities after the molecular-level mixing process thereby creating new properties.

Especially for optical device materials, desired properties of hybrid glasses can be chemically designed and then prepared by incorporating functional organic fragments between inorganic oxides.

With this strategy, novel optical device materials with beneficial properties can be obtained by embedding organic spacers into silicate network to create organically modified hybrid glasses as demonstrated in earlier publications (Figure 3). [1-10] Furthermore, the molecular-level mixing technique doesn't show any significant phase separation; because, the molecular-level hybridization is based on a microscopically homogeneous mixing and thus the uniform distribution of organic and inorganic moieties in a domain size at the molecular level is provided.

Those organically modified glasses mixed at the molecular scale, were provided by a molecular modification, which inserts desired organic fragments between two inorganic oxides to create entirely new optical properties. Figure 3 shows the molecular-level hybridization to produce polysilsesquioxanes.

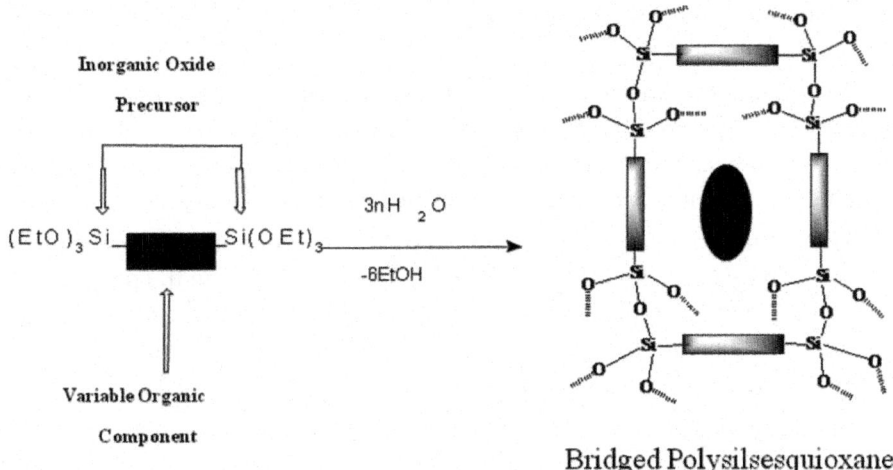

Figure 3: Molecular-level hybridization technique to produce polysilsesquioxanes. [6].

Bridged polysilsesquioxanes are a new family of molecular-level composites, which also a new version of hybrid glasses; these are prepared from hybrid sol–gel processable monomers through a sol-gel polymerization. These are microscopically homogeneous with uniform distribution of the organic spacers and thus also show an excellent optical transparency. There are many publications of novel optical device materials based on organic/inorganic hybrid glasses. [11-17]

In optical device materials, organically modified silicate glasses/transparent polymers have been actively pursued to develop novel optical devices, such as lasers, optical switches, optical fibers, waveguides, laser amplifiers, optical displays, and data storage devices.[11-17]

In addition, we can also control the porosity of those organically modified silica, polysilsesquioxanes, by inserting different molecules or sizes of organic spacers as shown the void space in Figure 3 (right); due to the insertion of organic spacers, the porosity of the resulting hybrid glass significantly increased. [18-26]

The expanded pores allow us to dope semiconductor or metal particles without any significant mechanical cracks. The molecular-level hybridization also solves the phase separation problem during molecular-level mixing process.

The pore size of organically modified silicate materials can be controlled by both choices of organic spacers and sol-gel conditions. Those molecularly designed hybrid glasses have shown high surface areas and a relatively narrow distribution of pore sizes that range from the high micropore to the low mesopore domain (15-100 Å).

Several organic/inorganic hybrid sol-gel monomers have been molecularly designed and then synthesized for developing novel hybrid glasses. A variety of bridged polysilsesquioxanes, organically modified hybrid silica, have been designed by the molecular-level hybridization (Figure 4). [18-26]

Those functional organic spacers inserted in the silicate networks (Figures 3 and 4) can serve as dopant precursors to growth particles in the porous glassy hosts. Those void volumes in the glassy hosts can be a matrix for the growth of quantum dots, such as semiconductor or metal particles. No phase separation occurs during the corporations of those dopants.

In Figure 5, those sol-gel processable monomers contain functional groups, which are sol-gel polymerized under either acidic or base condition, and then produce highly porous xerogels. [4]

Due to the highly porous silicate matrices, we doped various dopants without any phase separation; for example, we prepared highly nano-porous polysilsesquioxane systems, and then controlled sol-gel conditions to dope nano-sized transition metal particles or semiconductor particles, such as CdS

[18,19, 21], chromium [20, 21], iron [22], cobalt [27], and platinum [28] into the silicate hosts.

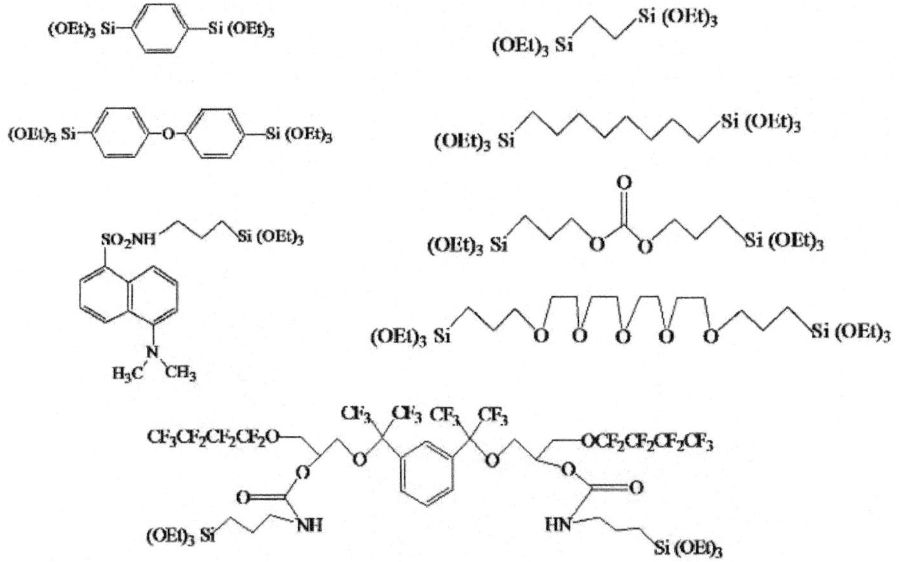

Figure 4: A variety of sol-gel processable monomers prepared by the molecular-level hybridization.

NOVEL OPTICAL DEVICE MATERIALS FOR LASER AMPLIFIER [25]

We have developed novel laser devices based on those organic/inorganic hybrid glasses prepared bythe molecular-level hybridization.

Rare-Earth Ion Doped Laser Amplifier

Silicate-based optical fibers and planar waveguide amplifiers are widely studied for optoelectronic applications because of the superior chemical resistance and compatibility with other optical devices based on polymeric materials.

Transparent silica doped with rare-earth metal ions has been used for laser amplifiers, such as photonic fiber amplifiers, solid-state lasers, compact laser amplifiers, ultra short pulse lasers, high-power lasers, so on. [29-39]

Planar waveguides and fibers doped with rare-earth metal ions are a key challenge; thus, an enormous amount of publications in rare-earth metal ion doped optical devices has been found. [29-37] Fabrication of optical devices with high resolution offers an efficient approach to minimize the cost and size of optical amplifiers.

High gain laser amplifiers can be achieved by improving several relative factors, such as optical losses, phonon energies, pumping powers and distances, fluorescence life times, and refractive indices of optical medium. The lasing efficiency can be also improved by changing the chemical environment of rare-earth metal ions.

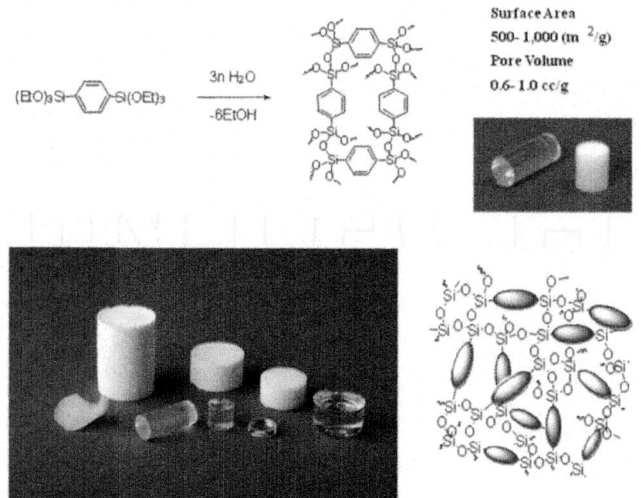

Figure 5: Highly porous polysilsesquioxanes. [4].

In a laser amplification based on rare-earth metal ions, erbium (Er) has been widely used as a gain medium due to its strong fluorescence at 1540 nm, which is a useful wavelength in optical amplifications. Especially, erbium-doped fiber amplifiers (EDFA) dominate this object of high gain optical amplifications. [32-35]

Since the performance of laser amplifiers is significantly influenced by optical media, scientists have been investigating organic/inorganic hybrid silicate hosts doped with Er^{3+} ions, in order to achieve high NIR efficiency and low phonon energy of the matrix to shorten the pumping distance and thus to obtain proper gain/life time. [32-35, 38, 39]

However, inorganic based optical media have shown a limitation to adjust the chemical environment of doped metal ions. For example, conventional silicate-based laser amplifiers often fail to produce high lasing performance because of a strong absorption raised from the OH-group at 1540 nm. The low concentration of erbium-ions in silicate host and the small absorption cross-section of the erbium-ions also limit the performance of normal silicate-based laser amplifiers since the doping level of rare-earth ions in glassy hosts significantly depends upon lasing efficiency.

Organic/inorganic hybrid glasses have been actively pursued as an alternative to conventional silicate glass for fabricating laser devices due to low temperature process and the promise of bringing new optical properties that are not possible from inorganic silica. [1, 40-42]

Furthermore, organic/inorganic hybrid silica is a good candidate to adjust the fluorescence environment of rare-earth metal ions by incorporating desired organic precursors into glassy hosts without any phase separation; usually, in inorganic silicate hosts, rare-earth ions tend to aggregate due to the absence of non-bridging oxygens, which cause a significant deduction of lasing efficiency. The synthesis of chemically modified silica with homogeneous doping of rare-earth ions is a key contributor to improve the performance of laser amplifications.

Optical device materials are required good optical transparency, controllable porosity, chemical purity, tunable refractive index, so on. Our goal for achieving those desired properties is to improve the fluorescence environment of Er^{3+} -ions in glassy hosts. For that, we devoted our attention to achieve an excellent chemical homogeneity of Er^{3+}-ion environment in glassy hosts.

DESIGN OF FLUOROALKYLENE-BRIDGED XEROGEL DOPED WITH ER $^{3+}$/CDSE

To demonstrate enhanced performance in laser amplifiers, we designed fluoroalkylene-bridged polysilsesquioxanes doped with Er^{3+}/CdSe nano-particles. [25] The fluoroalkylene-bridged silica was initially designed to reduce the phonon energy of the glassy host. Furthermore, CdSe nano-particles were also provided for further manipulation of the photochemical environment of erbium-ions in the matrix.

Use of organic silanes incorporates an organic fragment as an internal component of the silicate network. Sol–gel polymerization involves the hydrolysis of ethoxysilyl groups to yield silanols follows by subsequent condensation to form siloxane (Si–O–Si) linkages. In the sol-state, the condensation is insufficient to form a network, and the solution remains processable.

When sufficient cross-linking occurs, a network is formed and the transition from the sol-state to the gel state occurs. The presence of organic fragments within the 3-D structure imparts organic character to the hybrid glass that changes the microenvironment of additional fluorescence rare-earth ions incorporated in the glassy host.

In this study, fluorocarbon-linkages were designed to achieve high hydrophobicity within the hybrid glassy matrix. Erbium isopropoxide was also employed as the source for the Er^{3+} ions. Furthermore, CdSe nano-particles were also prepared and incorporated into the fluorinated glassy matrix to reduce the phonon energy of the glassy host (the phonon energy of CdSe = 200 cm^{-1}). [43] In principle, when rare-earth metal ions are excited in transparent glassy matrices, they can behave as a laser, which enables amplification of the incident light intensity. The lasing performance significantly relies on rare-earth metal ion doping level, host materials' physicochemical property, and chemical homogeneity.

Experimental

A set of three different sol–gel processable monomers were prepared; tetraethoxysilane (TEOS), 1,6-bis (triethoxysilyl)hexane, and 2,2,3,3,4,4,5,5-octafluoro-1,6-hexanediol bis(3-triethoxysilyl)propyl carbamate.

Tetraethoxysilane (Teos)

TEOS was purified by drying over 4 Å molecular sieves followed by a vacuum distillation.

Synthesis Of 1,6-Bis(Triethoxysilyl)Hexane

1,6-Bis(triethoxysilyl)hexane was synthesized by a 'hydrosilylation' of the corresponding α, ω-alkyldienes with triethoxysilane employing chloroplatinic acid (H_2PtCl_6) as a catalyst. 1,5-Hexadiene (12.3 g, 0.15 mol), triethoxysilane (54.1 g, 0.33 mol), and chloroplatinic acid (1mL of 7.5×10^{-5} mol in isopropanol) were placed in a round bottle flask. After 10 hours of simple stirring process at a room temperature, the reaction mixture darkened.

The reaction was monitored by GC. The crude product was purified by a vacuum distillation with a resulting purity of 99.87 % by GC analysis: bp 130 ∘C/0.1 mmHg. The final product was verified by NMR analysis and mass spectroscopic analysis and was consistent with data previously report. [11]

Synthesis Of 2,2,3,3,4,4,5,5-Octafluoro-1,6-Hexanediol Bis(3-Triethoxysilyl)Propyl Carbamate

2,2,3,3,4,4,5,5-Octafluoro-1,6-hexanediol (1g, 3.8 mmol) and 3-isocyanatopropyl triethoxy-silane (1.9 g, 4.3 mmol) were placed in a round bottom flask with a magnetic stirrer. The flask was sealed, purged with nitrogen, and 10 mg of dibutyl-tin-dilaurate was injected into the vial using a syringe as a catalyst.

The reaction was kept at room temperature under a nitrogen flow for several hours and monitored for the disappearance of the isocyanate peak at $2270 \ cm^{-1}$ (CN) in FT-IR. As the isocyanate group was converted to urethane group, identical peaks of $3400 \ cm^{-1}$ (NH) and $1720 \ cm^{-1}$ (CO) were observed. The product was dissolved in methanol and the tin catalyst was completely removed using a separated funnel. Methanol was then removed by a rotary evaporator. The product was used without further purification.

Er^{3+}-Ion/Cdse Doping Procedure

A sol-gel processible monomer, erbium isopropoxide (Chemat Technology), was used as a source of Er^{3+}-ions. CdSe nano-particles were synthesized by a previously reported procedure. [44, 45] Those dopants were incorporated by mixing with the appropriate sol–gel mixtures (Table 1); xerogels-a5 and -a10 denote higher erbium concentrations (5 and 10 times higher) than that of xerogel-a system.

We also examined the homogeneity of three sol–gel mixtures (Figure 6). Ethanol was used as a solvent. A visual inspection was carried out to determine the comparative homogeneity of those sol–gel mixtures containing erbium isopropoxide (Figure 6-1) or both of erbium isopropoxide/CdSe nano-particles (Figure 6-2). In a mixing test (Figure 6), erbium isopropoxide was indicated as a bright pink color in the photographs. The CdSe nano-particles also show a characteristic bright orange-color in those mixtures.

Table 1: List of xerogels doped with different components

Xerogels	Silicate matrices	Dopants
T-Xerogel	TEOS	None
H-Xerogel	Hexylene-	None
F-Xerogel	Fluoroalkylene-	None
Xerogel-a	TEOS	Erbium ions
Xerogel-b	TEOS	Erbium ions/CdSe
H-Xerogel-a	Hexylene-	Erbium ions
H-Xerogel-b	Hexylene-	Erbium ions/CdSe
F-Xerogels-a, -a5, and -a10	Fluoroalkylene-	Erbium ions
F-Xerogels-b, -b5, and -b10	Fluoroalkylene-	Erbium ions/CdSe

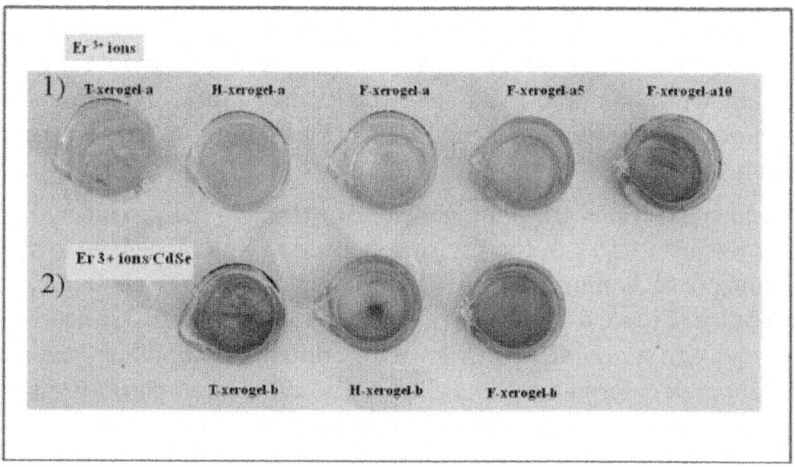

Figure 6: Mixing test.

Sol–Gel Procedures

Those sol–gel mixtures were then polymerized to produce the condensed xerogels under acidic condition using HCl as a sol-gel catalyst. Those xerogels were kept in a vacuum oven for 1–2 days to remove the remaining solvent and complete condensation.

Solid State Nmr Experiments

The ^{29}Si solid state NMR analysis of condensed xerogels was performed using a Varian Unity 400 solid state NMR spectrometer. The degree of condensation of undoped xerogels was computed from the single pulse magic angle spinning (SP/MAS) technique. A line fitting routine was also used in the analysis of the ^{29}Si NMR resonance in each spectrum to establish the siloxane ratio in the different structures.

Fluorescence Measurements

Fluorescence study of Er^{3+}-ions doped into those glassy matrices was carried out by using the Ar+ ion laser (488 nm). Laser power densities ranging from 1.5 to 3 Wcm^{-2} were used for the measurements.

Results And Discussions

Mixing Test

During the mixing test (Figure 6), we observed that the TEOS-based mixtures revealed a substantial degree of undesirable phase separations after doping with the Er^{+3}-ion sources or Er^{+3}/CdSe nano-particles. For example, in Figure 6-1, the T-xerogel-a mixture shows a significant phase separation even at the lower erbium concentration.

In contrast, hybrid sol–gel monomer mixtures showed significantly less phase separations (Figure 6). Hybrid sol–gel monomer mixtures accommodate and homogeneously distribute the Er^{3+}/CdSe source without any significant phase separation. In mixing test with hybrid sol-gel monomer mixtures, we observed no momentous phase separation, especially in the highly fluorinated sol-gel mixtures.

It is apparent that the TEOS-based sol–gel mixture has a rather limited solubility of erbium-ions, and hence a limited capability for fluorescence enhancement.

Subsequently, we provided CdSe nano-particles by following the earlier method [44, 45], and then added CdSe nano-particles into those sol–gel monomer mixtures containing the the Er^{3+}/CdSe source. Usually, CdSe nano-particles synthesized in colloidal configuration are suitable for incorporation into a variety of hosts including sol–gel mixtures. The comparative homogeneity of the three sol-gel monomer mixtures containing both of the erbium isopropoxide and CdSe nano-particles is also shown in Figure 6-2.

In Figure 6-2, TEOS-based mixture (T-xerogel-b) shows the CdSe nano-particles segregated in the mixture; the orange-colored CdSe nano-particles are observed to phase separate within the mixture.

In contrast, a hybrid sol-gel monomer system (H-xerogel-b) shown in Figure 6-2, the CdSe nano-particles mixed better than the T-xerogel-b. In the H-xerogel-b mixture, most of CdSe particles were dissolved, except some of undissolved orange-colored CdSe residues toward the middle of the container. In fluorinated mixture (F-xerogel-b) shown in Figure 6-2, CdSe nano-particles were incorporated without phase separation. This result demonstrates that the fluoroalkelene-bridged sol-gel monomer has the capability of uniformly incorporating both types of dopants without any phase separation.

Solid State Nmr Analysis

The chemical composition and the degree of condensation for those condensed xerogels can be determined by solid state nuclear magnetic resonance, infrared, and Raman spectroscopies. [1]

We employed a solid state NMR analysis to determine the degree of condensation of hybrid glassy hosts. ^{29}Si solid state NMR was used to identify the Si–O–Si bonds in variety states of condensation for three matrices. Single pulse magic angle spinning NMR methods were employed for the characterization of T-xerogel, H-xerogel, and F-xerogel to calculate the degree of condensation.

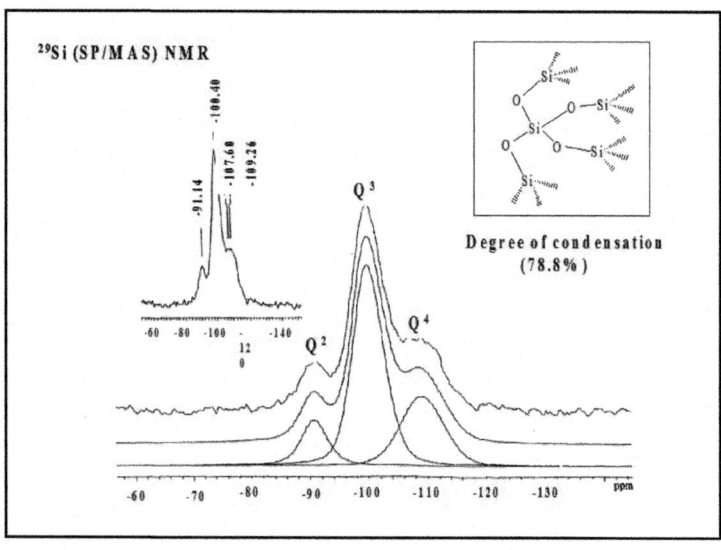

Figure 7: Si solid state NMR spectrum for undoped T-xerogel.

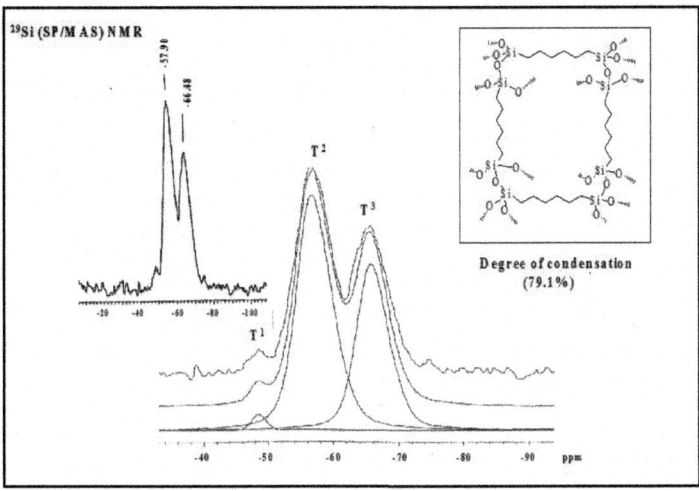

Figure 8: Si solid state NMR spectrum for undoped H-xerogel.

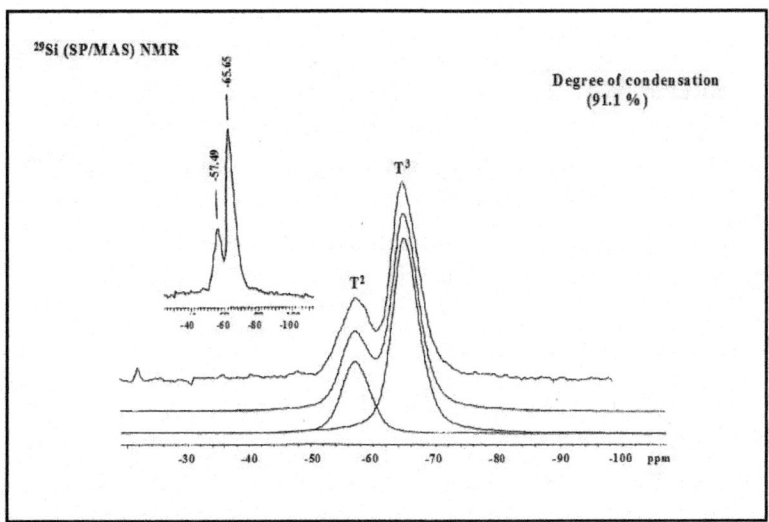

Figure 9: Si solid state NMR spectrum for undoped F-xerogel.

Figures 7-9 show the result of ^{29}Si SP/MAS solid state NMR analyses and peak deconvolution lines of both normal and modified silicate systems.

Figure 7 shows the ^{29}Si SP/MAS solid state NMR spectrum of the undoped T-xerogel. As shown inFigure 7, it reveals three peaks, which correspond to the Q^2, Q^3, and Q^4. The degree of condensation for the T-xerogel was calculated to be 78.8%.

Figure 8 shows the ^{29}Si SP/MAS solid state NMR spectrum of the undoped H-xerogel. Three peaks shown in spectrum correspond to the T^1, T^2, and T^3. The degree of condensation of the H-xerogel was also computed to be 79.1%.

The ^{29}Si SP/MAS solid state NMR spectrum of the undoped F-xerogel is given in Figure 9. The T^1 peak is not observed in this system. Also, the T^3 peak intensity in Figure 9 is higher than that of T^2. We found that the degree of condensation was dramatically increased to 91.1% in this case.

The high degree of condensation in the undoped F-xerogel can be explained as a result of the electron-withdrawing effect of the fluorine, which causes a partial positive charge at the silicon facilitating nucleophilic attack in the sol–gel process thus accelerating hydrolysis and condensation.

The high degree of condensation, disappearance of T^1 peak, and enhanced T^3 peak in the F-xerogel indicate a low level of hydroxyl group content and a greater degree of condensation when fluorinated alkylene groups are present in the glassy hosts.

Si–OH groups show a high phonon energy (3000–3500 cm^{-1}) at 1540 nm. [43]

The reduction in hydroxyl content in the F-xerogel matrix decreases the phonon energy of the matrix. The exclusion of moisture from the high hydrophobicity of the fluoro-alkylene groups in the F-xerogel matrix may also contribute to reduce the absorption at 1540 nm with a concomitant increase the fluorescence intensity of Er^{3+}-ions.

All these effects contribute to the increased fluorescence from erbium-ions in the fluorinated hybrid glassy matrix.

Xps Analysis

We also employed a XPS study to determine the chemical composition of F-xerogel-a. [46]

A full XPS scan was obtained in the 0–1100 eV range. Detail scan was also recorded for the Er (4d) region. Figure 10 shows a full spectrum for F-xerogel-a. The XPS spectrum of F-xerogel-a shows an erbium peak at ~169 eV, which corresponds to the presence of Er$_2$O$_3$.

The atomic compositions were evaluated in this study. The concentration of each element (atomic %) was calculated; O(1s)—22.74 atomic %, C(1s)—53.29 atomic %, Er(4d)—1.49 atomic %, F(1s)—9.02 atomic %, Si(2p)—10.91 atomic %, N(1s)—2.55 atomic %. From the XPS analysis, it was estimated that the F-xerogel-a contains ~1.49 atomic % of erbium.

Fluorescence Measurements

Erbium-ions incorporated into glassy matrices exhibit well defined energy level transitions in 4f-shell electronic configurations.

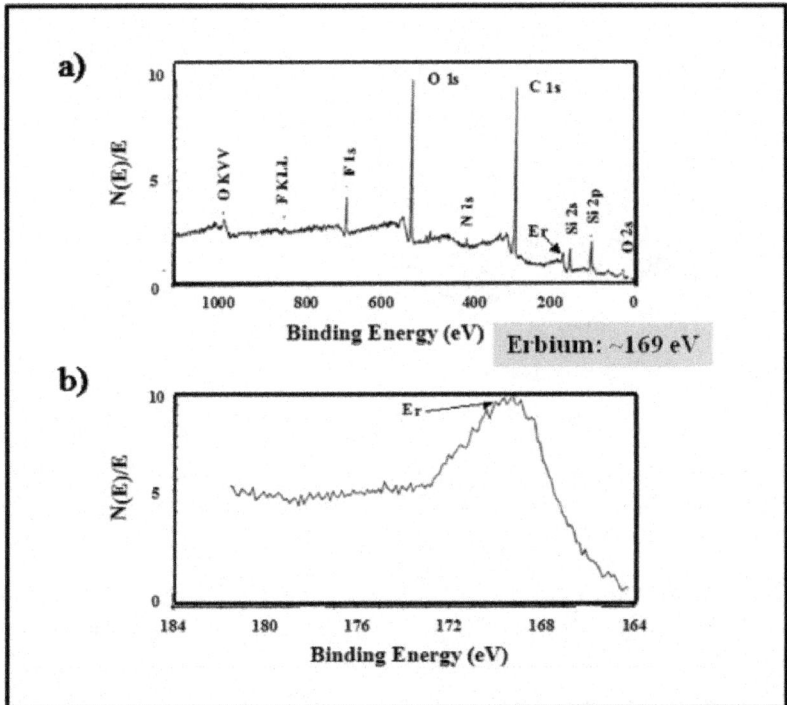

Figure 10: XPS Analysis of F-xerogel-a.

For erbium-ions, the $4I_{13/2}$ to $4I_{15/2}$ transition is important in optical communications; because, it results in fluorescence at 1540 nm, which is the most important wavelength regime for optical communication applications, especially in long-distance telecommunication networks. [47]

We examined how the fluorescence intensity of erbium-ions was dependent upon the matrix environment when fluorine and CdSe nano-particles were incorporated into hybrid glassy hosts. We carried out fluorescence analysis of erbium-ions around 1540 nm. The results are shown in Figures 11-13.

A comparison of fluorescence intensities from erbium-ions doped into different glassy hosts is shown in Figure 11 (H-xerogel-a and F-xerogel-a).

In Figure 11, fluorescence intensity of erbium-ions increased significantly more in F-xerogel-a than the other hybrid system of H-xerogel-a.

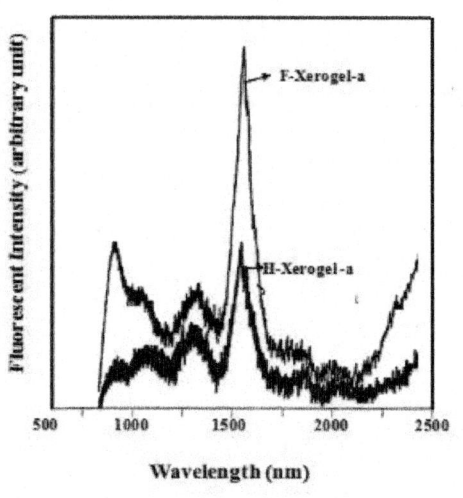

F-xerogel-a :   Fluoroalkyl-based xerogel
                with erbium-ion dopant

H-xerogel-a :   Hexylene-based xerogel
                with erbium-ion dopant

Figure 11: Photoluminescence of erbium-ions; different silicate matrices. Comparative fluorescence intensities of two different hybrid xerogels doped with Er^{+3} ions using a low power of 1.5 Wcm^{-2}.

The explanation of Figure 11 is that the enhanced fluorescence intensity of the fluorinated matrix is attributed to mainly its high hydrophobicity combined with the lower OH-group contents, which revealed in NMR study in Figure 9. Solid state silicon NMR analysis indicates an enhanced condensation in fluorinated xerogel compared to that of alkylene-bridged xerogel. The fluorinated hosts also showed an excellent chemical homogeneity in mixing test (Figure 6), which significantly affects the lasing performance.

Further investigations in fluorescence studies have been carried out. Since it is important to consider the erbium-ion concentration effect, fluoroalkylene-based glasses doped with two different levels of erbium concentrations were prepared for the determination of erbium-concentration effect (Figure 12).

Intensity of those xerogels (F-xerogel-a and F-xerogel-a5) was measured at a power density of 3 Wcm^{-2}. As shown in Figure 12, the fluorescence intensity increases (the upper curve) in the case of F-xerogel-a. It can be explained that the higher erbium concentration of F-xerogel-a5 caused a low lasing efficiency due to the "self-quenching effect."

Fluoroalkylene-bridged xerogel containing Er^{3+}-ions shows significantly reduced absorptions at the 1540 nm by reducing amounts of uncondensed hydroxyl groups.

The presence of CdSe nano-particles also significantly influences the fluorescence environment of Er^{3+}-ions in different glassy hosts, resulting in the increased fluorescence intensity. [43].

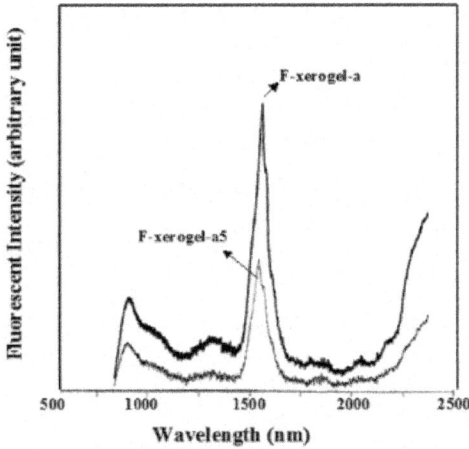

F-xerogel-a: Fluoroalkyl-based xerogel with Erbium ion/LOW Concentration (1.49 atomic %)

F-xerogel-a5: Fluoroalkyl-based xerogel With Erbium ion/HIGH Concentration (5 x 1.49 atomic %)

Figure 12: Photoluminescence of erbium-ions; different erbium concentrations. Comparative fluorescence intensities of xerogels with different Er^{3+}-ion concentrations (F-xerogels-a and -a5) using a power of 3.0 Wcm^{-2}.

CdSe nano-particles were used to modify the photochemical environments of erbium-ions in glassy hosts by taking advantage of a low phonon energy of CdSe phase (200 cm^{-1}) [43], since the incorporation of semiconductor nano-particles resulted in an enhancement of the semiconductor-to-erbium transfer when the quantum well and erbium-ion transition energies became close.

We thus examined the fluorescence of erbium-ions surrounded by CdSe nano-particles since it was anticipated that the presence of CdSe in the modified glassy hosts would affect the fluorescence performance (Figure 13). In order to test this, we prepared fluoroalkylene-bridged hybrid glasses doped without and with the CdSe nano-particles, F-xerogels-a5 and F-xerogel-b5, respectively.

As shown in Figure 13, the fluorescence intensity of F-xerogel-b5 is dramatically increased, which indicates an improvement in lasing efficiency by modifying photochemical environments of erbium-ions.

By taking advantage of the structural features and uniform doping capability in modified glassy matrices, we successfully demonstrate that the fluorescence environments of erbium-ions can be a key to improving the performances of optical devices like laser amplifiers to overcome the limitations in inorganic silica.

In conclusion, we have demonstrated here a promising result in laser amplifications by employing bridged polysilsesquioxanes doped with Er^{3+}-ions/CdSe nanoparticles.

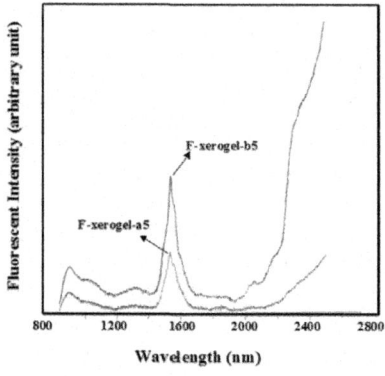

F-xerogel-a5: Fluoroalkyl-based xerogel with Erbium-ion dopant

F-xerogel-b5: Fluoroalkyl-based xerogel with Erbium-ion and CdSe dopant

Figure 13: Photoluminescence of erbium-ions; CdSe nanoparticle effect. Comparative fluorescence intensities of fluoroalkylene-bridged xerogels doped without and with the CdSe nano-particles (F-xerogels-a5 and -b5), respectively, using a high power of $3\,Wcm^{-2}$.

To avoid the high phonon energy raised from OH-groups ($3000-5300\ cm^{-1}$) at 1540 nm [43], we designed highly fluorinated hybrid glasses to shorten the fluorescence-level life times of dopants, which adversely affect optical device performance.

The presence of CdSe nano-particles, by virtue of its lowering of phonon energy, also appears to significantly influence the nature of the surrounding photochemical environment of Er^{3+}-ions in the fluorescence study.

From those study, we found that the control of such optical materials' properties affect the performance of optical devices via molecular tailoring strategy, which is molecular-level hybridization technique.

NOVEL OPTICAL DEVICE MATERIALS FOR ACOUSTIC WAVE [26]

Novel Optical Device Materials Based On Polysilsesquioxanes

The preparation of semiconductors, metals, and ions in a variety of transparent materials has been actively pursued for optical devices. Scientists have taken a great attention to incorporate metal particles/ions in glassy hosts to develop high performance optical devices.

Chemists also have taken intensive challenges on how to achieve homogeneous incorporations of semiconductors or metal particles in glassy hosts, which are significantly influenced by diffusion of reagents, the number of nucleation sites, stabilization of growing particles by surface functionality, the boundary constraints of the growth matrix, and the opportunity for equilibration or "ripening" of particles formed under kinetic growth conditions. [2, 21, 48, 49]

Organically modified hybrid glasses, which were prepared by 'molecular-level hybridization' (Figure 3), are a good candidate to develop laser device materials due to their easy processability and chemical modification.

Polysilsesquioxanes can be prepared by 'molecular-level hybridization', which inserts different types and sizes of organic spacers between two inorganic oxides. The sol-gel chemistry was employed to covalently incorporate semiconductors/metals in various oxidation states as an integral component of hybrid silicate matrices.

The 'molecular-level corporation techniques' also can be employed for the incorporation of semiconductors or transition metals/ions dispersed in optically transparent silica to prepare novel optical devices.

Especially, polysilsesquioxanes are microscopically homogeneous due to uniform distribution of organic and inorganic moieties in a domain size at the molecular level. Hence, these hybrid glasses can be used for optical device materials since the molecular-scale mixing process significantly reduces phase separation and thus produces high quality of optical clearance.

In this work, we molecularly designed a novel hybrid glass to develop alkylene-bridged polysilsesquioxanes doped with $Cr^o/CrOx$ phases (Figure 14). The doped xerogel film effectively generates a HUGE ACOUSTIC WAVE.

Periodic Alignment Of Alkylene-Spacers To Create Molecular-Scale Optical Grating For Acoustic Wave

Usually, scientists prepare 'periodic metal frames' to create 'diffraction grating' at the bulk-scales. For example, in a spectroscopic monochromator, 'optical grating effect' is used to split, and then diffract the light into several beams traveling in different directions.

In 'grating equation', the directions of these beams rely on the spacing/ distances of the grating, which has a periodic structure of the media, and the wavelength of the light. Gratings, which modulate the phase rather than the amplitude of the incident light, can be also produced. A key idea of the periodic alignment of alkylene-spacers is to create 'a molecular-scale diffraction grating effect' and thus to generate a huge acoustic wave.

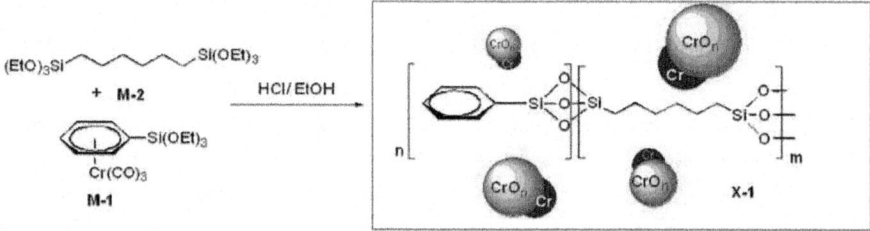

Figure 14: Hexylene-bridged polysilsesquioxane doped with Cr°/CrOx.

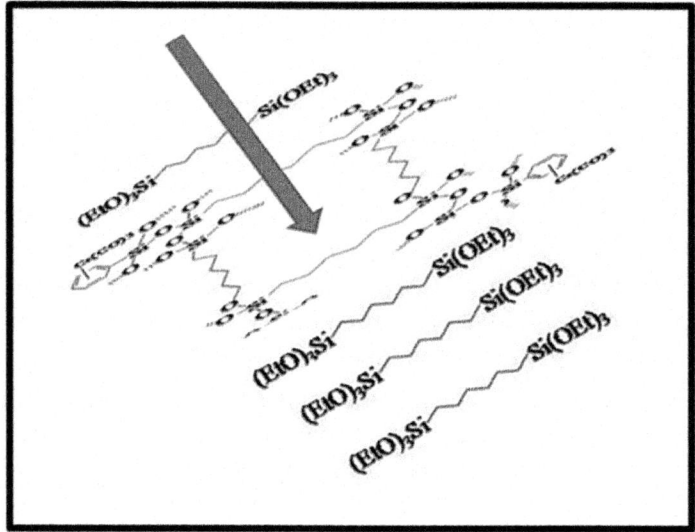

Figure 15: Periodic alignment of alkyl-chains to create "optical grating" at the molecular scales.

Figure 15 illustrates a light that passes through 'optical grating at the molecular-scales' designed in the hybrid silicate host. As you can see, the periodic alkylene-spacers create 'effective diffraction grating at the molecular scales' and thus produces a huge acoustic wave. This is a new optical phenomenon, which hitherto hasn't been discovered.

In addition, the distance between those alkyl-spacers can be also "controlled" by both choices of different organic spacers and sizes of sol-gel monomers. This is an effective method of developing new laser device materials based on organically modified silica, polysilsesquioxanes.

Usually, when the laser beam goes through a solid medium, the density wave is linear; because, in solid media, heat doesn't decay through the solid medium effectively.

Interestingly, the highly compressed xerogel, hexylene-bridged polysilsesquioxane doped with $Cr^o/CrOx$ phases showed a strong 'acoustic response' as much as a liquid.

In our laser experiments, we calculated physical parameters of hexylene-bridged xerogel doped with $Cr^o/CrOx$ phases. The coefficient of phonon diffraction (D) of the doped xerogel was FIVE times smaller than that of normal glass. Which means that the thermal conductivity of the doped xerogel, is FIVE times less than that of normal glass.

In addition, the diffraction efficiency, absorption light efficiency, (45%) of the doped xerogel is higher than that of methanol (25%), which means the COMPRESSIBILITY of the doped xerogel is as effective as liquid. The acoustic refractive intensity and acoustic response generated from the doped xerogel was as strong as liquid. Therefore, the doped xerogel serves as a 'heat generator,' and thus the heat gets transferred into expansion or compression wave (acoustic waves) effectively.

There were a lot of efforts to develop novel laser device materials based on organic and inorganic hybrid silica. [50-54] Especially, this is a unique approach to create an effective optical grating, which brings a HUGE ACOUSTIC RESPONSE by setting up a molecular-scale optical grating effect. By building up the periodic structure of long alkyl-chains in the hybrid glassy hosts, we obtained a HUGE ACOUSTIC WAVE, which hasn't been found. This is a new optical phenomenon.

Novel Sol-Gel Condition To Produce Thin Films Of Alkylene-Bridged Xerogel Doped With Cr°/Crox

The new optical property partially rose from a new sol-gel mixing condition, which effectively produces HIGHLY COMPRESSED, THIN xerogel films.

Conventional sol-gel conditions often result in poor optical transparency/ mechanical property. For example, xerogels obtained from the conventional sol-gel conditions are thick and brittle materials. Those thick xerogels are difficult to handle, especially for optical applications.

The limitation motivated us to find a novel sol-gel mixing condition, which produces a highly compressed, thin sol-gel film with a smooth surface and excellent optical clarity.

We discovered a novel sol-gel mixing condition, which results in such a thin xerogel film of alkylene-bridged polysilsesquioxanes doped with Cr^{o}/ CrOx with a low thermal conductivity and high compressibility (Figure 16). [26]

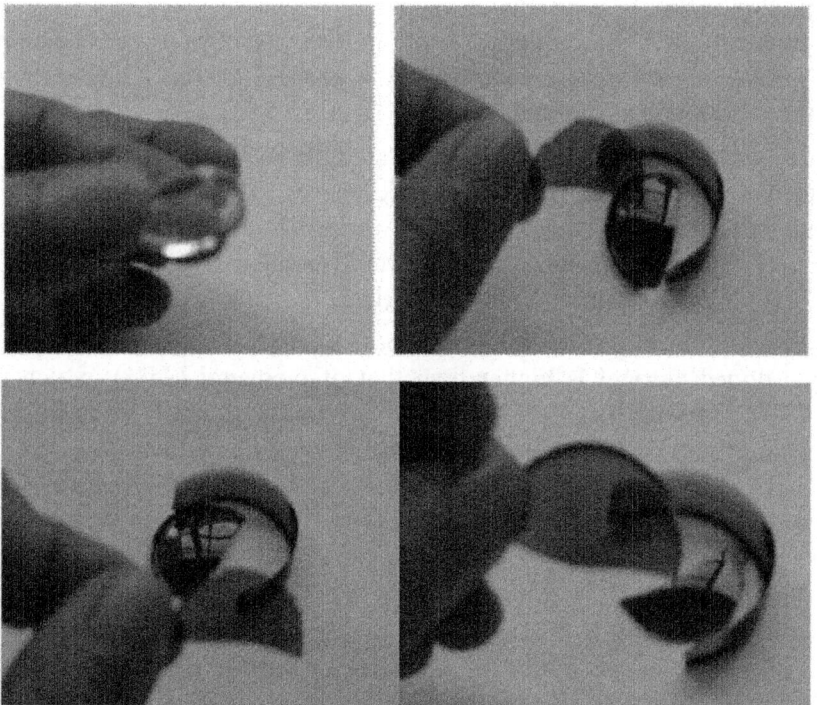

Figure 16: A comparative photos of undoped- (top left) and the Cr^{o}/CrOx doped- (top right and bottom) xerogels based on hexylene-bridged polysilsesquioxane under an acidic condition.

During the sol-gel polymerization, a volume of the sol-gel mixture was dramatically reduced, and then left a green sol-gel film with an excellent optical clarity (Figure 16).

For a source of chromium, we have synthesized a sol-gel processable chromium precursor [20, 26]; we prepared a green sol-gel film based on hexylene-bridged polysilsesquioxanes doped with $Cr^o/CrOx$ phases using the chromium precursor. [20, 26]

In Figure 16 (top left), it shows a undoped hexylene-bridged polysilsesquioxane prepared from our novel sol-gel mixing condition; it was plastic-like a xerogel monolith. As you see in Figure 16 (top, left), it shows an excellent optical transparency.

Doped xerogels were also prepared in Figure 16 (top, Right and bottom). Those green xerogels were obtained as "plastic-like thin films" with an excellent optical compressibility and low thermal conductivity. It was prepared without any mechanical damage/cracking problem.

Experiments

Preparation Of Doped Xerogels

Syntheses of sol-gel monomers, 1,6-bis(triethoxsilyl)hexane and chromium precursor, η^6-chromium tricarbonyl(triethoxysilyl)benzene, were carried out by following the earlier procedures, respectively. [11, 20]

Subsequently, hexylene-bridged xerogel doped with $Cr^o/CrOx$ phases was prepared by a copolymerization of η^6-chromium tri-carbonyl(triethoxysilyl) benzene and 1,6-bis(triethox-silyl)hexane under an acidic condition (Figure 17). A green glass was provided after the sol-gel polymerization (Figure 16).

Density of the doped xerogel was measured to be 1.2696 g/cm^3 using a He-gas pycnometer. From AAS analysis, the chromium amount was also analyzed to be 1.4% by weight.

Tem, Edax, And Electron Diffraction Analysis

A novel periodic alignment of alkylene-spacers in hybrid glass was specifically designed for creating 'a molecular-scale diffraction grating.' We employed TEM analysis to identify the molecular alignment built up in the hybrid silicate matrix.

The doped hybrid glass under an acidic condition (Figure 17) was ground to powders with a particle size (<150 μm), and deposited on a plasma-etched carbon substrate supported copper grid. TEM dark-field images were obtained using a Philips transmission electron microscope (TEM: CM 20/STEM, PW 6060).

The energy-dispersive X-ray diffraction (EDAX) pattern of the glassy particles was also obtained by an EDAX analyzer (Philips TEM-EDAX, PV 9800). For the electron diffraction, the camera length was calibrated experimentally with a gold standard, and an X-ray spectrum analyzer at 200 kV was used.

LASER ANALYSIS

To establish the optical properties of doped xerogels, we prepared a sample (<1mm thickness) fabricated with η^6-chromium tricarbonyl(triethoxysilyl) benzene loading of 2.4 % under an acidic sol-gel condition. The thin xerogel film exhibited a nonlinear property. The nonlinear optical (NLO) properties of doped xerogel films were measured by the degenerated into four wave mixing (DFWM) technique. [55] A quartz sample was used as a reference.

We used two types of lasers, a YAG laser with 50 ps pulse-width at 532 nm and a dye laser with 150 fs pulse-width at 640 nm. The output of either of lasers is split into two strong pump beams and a weak probe. The delay between two pump pulses is set to zero to create interference patterns in the doped sol-gel film. Variable delay line on the probe beam allows to measure the decay time of the diffracted beam.

Results And Discussions

Figure 17 describes a sol-gel procedure for the preparations of hexylene-bridged polysilsesquioxanes doped with Cr^o/ CrOx.

The sol-gel process was carried out by a copolymerization of two sol-gel monomers, $\eta6$-chromium tricarbonyl(tri-ethoxysilyl)benzene (M-1) and 1,6-bis(triethoxsilyl)hexane (M-2) (Figure 14).

Those sol-gel monomers can be produced bridged Si-O-Si network under either acidic or basic condition, and thus processed into transparent glasses, glassy films, fibers, xerogels, and aerogels, and monoliths. [1,6] From the basic condition, it produced hybrid glasses containing chromium metal particles; because, the M-1 was stable under the basic sol-gel condition.

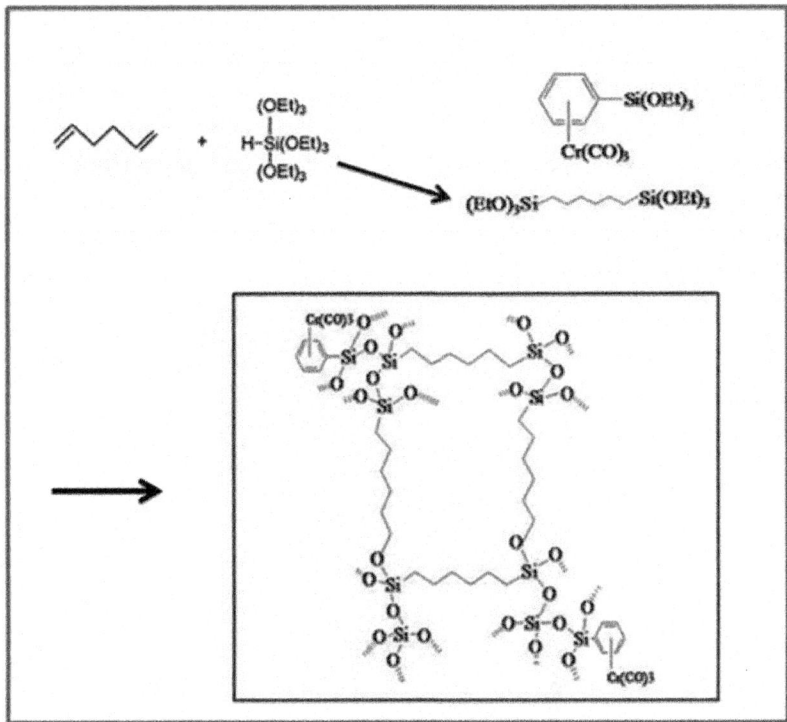

Figure 17: Synthesis of hexylene-bridged xerogel attached with the Cr° precursors.

In contrast, the acid-catalyzed system produced thinner sol-gel films than those of glasses obtained under base condition (Figure 16). Under the acidic sol-gel condition, decomposition of the M-1 competes with sol-gel copolymerization. The product of acid catalyzed decomposition reaction is "chromium oxides" and H_2 (Figure 17). [56-58]

In Figure 18, TEM images of alkylene-bridged silica doped with Cr°/CrOx phases reveal unusual nano-fringe patterns, which rose from the lattice fringes of the aligned alkyl-spacers in the silicate matrix. The novel molecular design results in 'a molecular level grating characteristic' when laser light passes by those periodic carbon-chains (Figure 15).

Based on those nano-fringe patterns, an effective optical grating was created in the hybrid silicate matrix. The TEM images of the doped hybrid glass reveal nano-fringe patterns, which are highly organized nano-periodicity (pointed with arrows in Figure 18). The nano-periodic patterns are sustained over substantial domains and appear to arise from lattice fringes.

In short, the formation mechanism of these nanoperiodic features observed in the TEM images arises from the highly arranged alkylene-spacers in the sol-gel monomer (Figure 15).

EDAX and electron diffraction analyses of these dark regions shown in the TEM images were also performed. In the EDAX spectrum, a Cr (Kα) peak was observed; thus, the dark contrast shown in the TEM images (Figure 18) was identified as a chromium phase spread over the hybrid glassy host by both of EDAX/electron diffraction analyses.

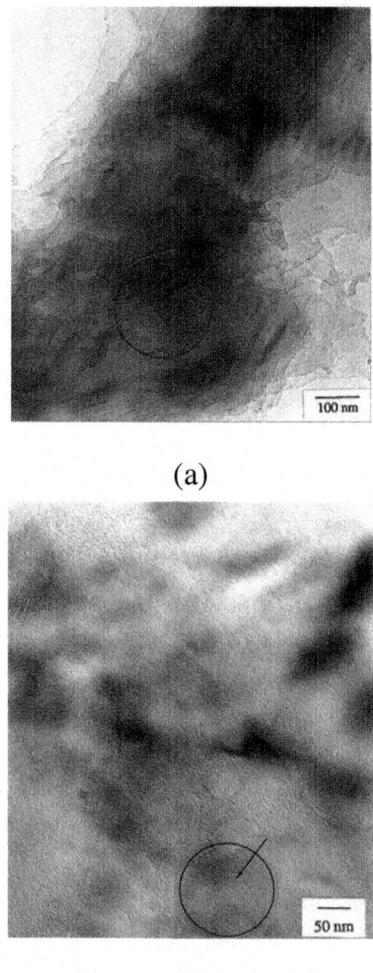

(a)

(b)

Figure 18: a and b. TEM Image of Cr°/CrOx-doped hexylene-bridged poly-silsesquioxane under acidic condition; it reveals unusual nano-fringe patterns with alternating features of a lattice spacing of about 50 Å in different areas.

The Cr°/CrOx phases were produced by a chemical reaction, which is a simultaneous sol-gel copolymerization then decomposition of the chromium precursor under the acidic condition. The electron diffraction pattern of the dark areas in TEM images was also identified as a mixture of Cr°/ CrOx phases.

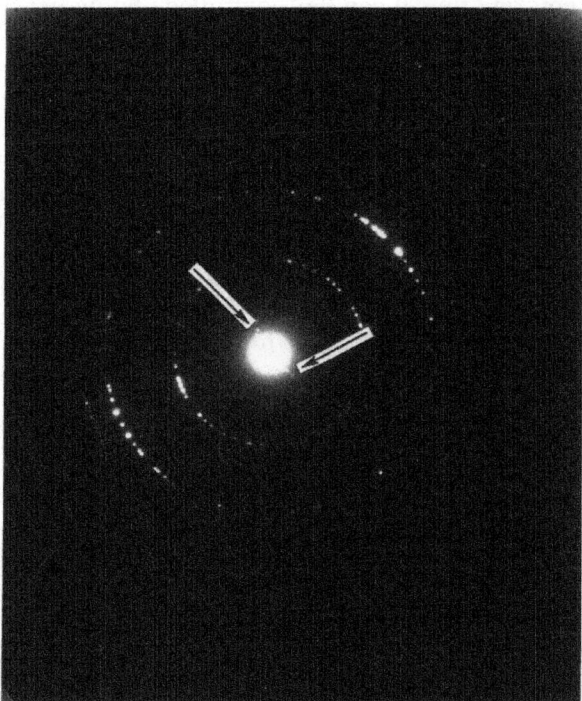

Figure 19: Electron diffraction pattern of doped hybrid glass; circled diffraction patterns correspond to the d-spacings of chromium metal. Highlighted features (arrows) correspond to the nanoperiodicity in the TEM images in Figure 18. From a distance between two diffraction spots point out each arrow a lattice spacing of the nanoscale fringes was calculated to be ~50 Å.

In order to calculate a lattice space of those highly organized nano-fringe patterns shown in the TEM images, we also carried out electron diffraction analysis of the doped sol-gel film. The result is shown in Figure 19. The nanoperiodicity gives rise to features in the electron diffraction pattern.

Figure 19 shows the circled diffraction patterns that a rise from crystalline chromium metal and a set of diffractions near the center of the beam corresponded to the nanofringes observed in TEM images (Figure 18).

From the diffraction pattern corresponded to the nano-fringes, a lattice space of the nanostriped patterns observed in TEM images was calculated about

50 Å from a distance between two diffraction spots in two sets of diffraction patterns, which are pointed with arrows in Figure 19; each set consists of two diffraction spots.

The distance between alkyl-spacers significantly relies on the optical grating efficiency, and thus it can be "controlled" by inserting different types and sizes of organic spacers.

As shown in Figure 14, the reactions occurring during the formation of the sol-gel xerogels are complex and include simultaneous sol-gel copolymerization and decomposition of M-1 under the acidic condition. At present, the mechanism of formation of these nanoperiodic features observed in the TEM images may arise from the highly arranged alkyl-spacers in the sol-gel monomer (M-2).

In this study, the nonlinear optical (NLO) properties of the doped xerogel film were measured by the degenerated four wave mixing (DFWM) technique. [12, 13, 55]

In femto- and pico-second experiments, "electronic χ^3" and "population χ^3" of the doped xerogel film have been measured (Figure 20). The DFWM signals for both "pure electronic and population χ^3" are shown in Figure 20 as a small spike around t = 0. It is asymmetric and longer than the laser pulse. The trailing edge of the peak has decay, which is probably connected with population relaxation.

In thermal nonlinearity, the coefficient of phonon diffraction, which is proportional to the coefficient of thermal conductivity, has been calculated from the DFWM experiments as 1.9×10^{-3} cm^2/sec using an equation, $D = \Lambda^2/4\pi\tau$ (where, Λ is the period of the diffraction grating and τ is the decay time of the thermal signal).

This number is about FIVE times lower than that of a normal glass. In other words, the thermal conductivity of the doped xerogel is FIVE times less than that of a normal glass. In acoustic study, the doped xerogel also shows an interesting new optical property, an effective generation of a large acoustic wave.

When the temperature of the doped xerogel at the maximum of the interference pattern goes up, the material expands then a counter propagating wave of expansion and compression start traveling inside the glassy host. Since the index of refraction depends on density of the material, on top of slow decaying thermal grating, we will have a dynamic diffraction grating propagating with the sound velocity.

By changing the delay on the laser probe beam, we measured the period of acoustic grating and extract the sound velocity of the material (Figure 20).

We used YAG laser at 532 nm and 50 ps pulse-width for a laser analysis for the doped xerogel obtained by the novel sol-gel condition, which produces thin xerogel films with unusually high compressibility (higher density).

Small spike around t= 0 corresponds to the signal due to electronic nonlinearlity (Figure 20). The time required for the acoustic wave to travel from one interference maximum to another is twice the time between t=0 and the peak of the acoustic signal. The sound velocity (C) in the thin doped xerogel was calculated from the distance between two acoustic waves ($\Delta\tau$) as $C = 3.2 \times 10^{5}$ cm/sec.

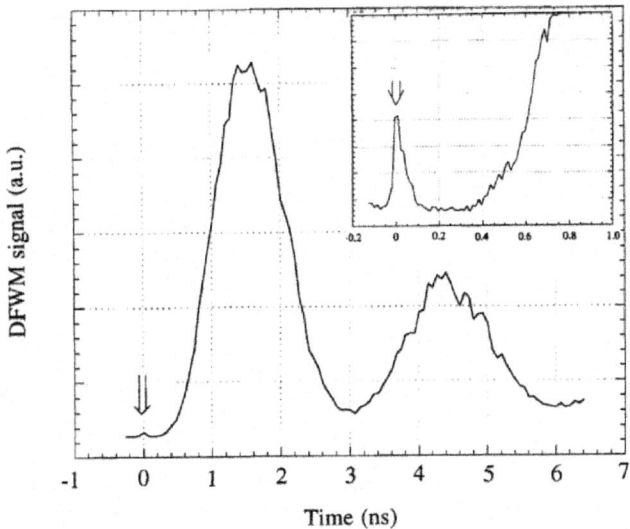

Figure 20: DFWM signals obtained from the doped xerogel measured in femto-second experiment (in a box) and pico-second experiment.

The signal decay time raised from the doped xerogel film was evaluated by the continuous wave (CW) probe experiment (Figure 21); it was obtained as 17 μsec.

Therefore, the doped xerogel serves as a heat generator in the slow nonlinearities due to the low thermal conductivity and high compressibility of the hybrid glass, thus the heat is transferred into expansion or compression wave (acoustic wave) very efficiently.

We also measured the diffraction efficiency of the probe beam at the delay time, corresponding to the peak of acoustic signal. At energy level about P = 0.47 J/cm², which is close to the optical damage threshold, the diffraction efficiency was 45 %.

The amplitude of the acoustic signal will depend upon how effectively the laser pulse energy is transferred to an expansion wave, which in turn depends on compressibility of the host materials. For a comparison we did the same measurement for a dye solution in methanol with the same optical density and the same energy density. The diffraction efficiency of methanol was 25 %, which means the compressibility of the doped xerogel film is as effective as liquid.

In a conclusion from the laser experiments, the doped xerogel film has a lower thermal conductivity than that of a normal glass. The compressibility of the doped xerogel is sufficiently high so the density grating formed in the doped xerogel could be effective to create high diffraction beam in the sound velocity.

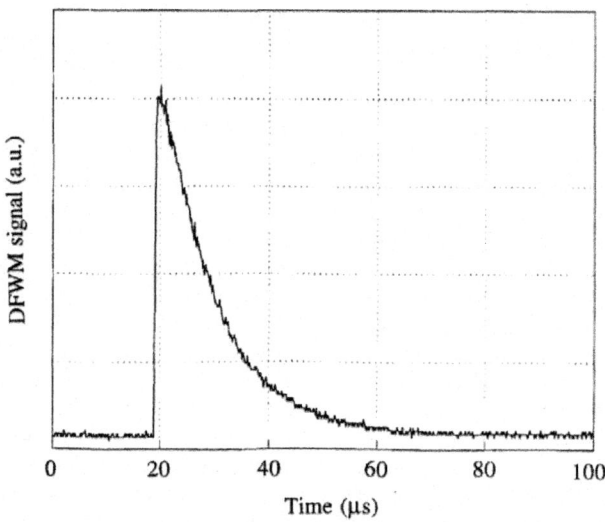

Figure 21: The decay of thermal signal for the doped xerogel measured in CW probe experiment.

We also believe that the nano-fringe patterns revealed in the TEM images (Figure 18), which rose from the lattice fringes of alkylene-bridged silicate matrix may result in an effective grating characteristics when the light passes by those long carbon-chains of alkylene-based polymeric networks.

The characteristics observed from the novel doped xerogel are new photonic properties, which hitherto have not been possible from simple physical mixing process of individual components at the bulk scale. In other words, those new optical properties are created by the molecular-level hybridization.

Based on these experiments, the (Cr°/CrOx)-doped polysilsesquioxane with low thermal conductivity and high compressibility are suggested as a new type of optical device materials for optical applications, for example diffraction beam modulators.

We demonstrated here some examples of creating new optical properties by designing novel molecular building blocks at the molecular-scales via the molecular-level hybridization.

CONCLUSIONS

We introduce the molecular-level hybridization for the preparation of polysilsesquioxanes, which are hybrids of inorganic oxides and organic network polymers.

Optically transparent hybrid glasses are prepared by a molecular tailoring technique, which produces sol-gel processible monomers.

The resulting hybrid glasses show novel optical properties, which would not be expected from those individuals by the loss of individuals' identities after the molecular-level mixing process, thereby creating entirely new properties.

REFERENCES

1.   J. Brinker, G. W. Scherer, 1990Sol-Gel Science: The Physics and Chemistry of Sol-Gel Processing: Published by Academic Press, Inc.

2.   K. J. Shea, D. A. Loy, O. W. Webster, 1989Chem. Mater., 1, 572.; Shea, K. J., Loy, D. A., & Webster, O. W. (1992) J. Am. Chem. Soc., 114, 6700.

3.   R. J. P. Corriu, J. J. E. Moreau, P. Thepot, C. M. M. Wong, 1992Chem. Mater., 4, 1217; Corriu, R. J. P., Thepot, P., & Mang, M. W. C. (1994) J. Mater. Chem., 4, 987.

4.   D. A. Loy, K. J. Shea, 1995Chem. Rev., 95 (5), 1431.

5.   G. Cerveau, R. J. P. Corriu, C. Lepeytre, 1997J. Orgmetal. Chem., 548 (1), 99.

6.   K. M. Choi, K. J. Shea, 1998Photonic Polymer Systems, Fundamentals, Methods, and Application: Edited by D. L. Wise et al. World Scientific Publishing Co. Pte. Ltd.

7.   K. J. Shea, D. A. Loy, 2001MRS Bulletin, 26, 368.

8.   D. A. Loy, K. . Rahimian, May, 2003Handbook of Organic-Inorganic Hybrid Materials and Nanocomposites: Edited by Hari Singh Nalwa Formerly of Hitachi Research Laboratory, Hitachi Ltd.

9.   K. J. Shea, J. Moreau, D. A. Loy, R. J. P. Corriu, B. Bour, 2003Hybrid Materials: Published by Wiley Inter science, New York.

10. L. Zhao, M. Vaupel, D. A. Loy, K. J. Shea, 2008Chem. Mater., 20, 1870.

11. H. W. Oviatt, Jr , K. J. Shea, J. H. Small, 1993Chem. Mater., 5, 943.

12. H. W. Oviatt, K. J. Shea, S. Kalluri, Y. Shi, W. H. Steier, L. R. Dalton, 1995Chem. Mater., 7, 493.

13. L. R. Dalton, 1998Polymers for Electro-Optic Modulator Waveguides in Electrical and Optical Polymer Systems; Fundamentals, Methods, and Applications: Edited by Wise, D. L., Copper, T. M., Gresser, J. D., Trantolo, D. J., & Wnek, G. E., (World Scientific, Singapore) Chapter 18.

14. Walcarius, 2001Chem. Mater., 13 (10), 3351.

15. Matsubara, 2003AIST Today, 3 (8), 32.

16. D. W. Schaefer, G. Beaucage, D. A. Loy, K. J. Shea, J. S. Lin, 2004Chem. Mater., 16, 1402.

17. Y. X. Pang, S. N. B. Hodgson, J. Koniarek, B. Weglinski, 2006J. of Magnetism and Magnetic Mater., 301(1), March 27, 83.

18. M. Choi, K. J. Shea, 1993Chem. Mater., 5, 1067.

19. M. Choi, K. J. Shea, 1994J. Phys. Chem., 98, 3207.

20. M. Choi, K. J. Shea, 1994J. Am. Chem. Soc., 116, 9052.

21. K. M. Choi, J. C. Hemminger, K. J. Shea, 1995J. Phys. Chem., 99, 4720.

22. K. M. Choi, K. J. Shea, 1995J. of Sol-Gel Sci. Tech., 5, 143.

23. K. M. Choi, J. A. Rogers, 2003J. Am. Chem. Soc., 125, 4060.

24. K. M. Choi, 2005J. Phys. Chem. B, 109, 21525.

25. K. M. Choi, 2007Mater. Chem. Phys., 103, 176.

26. K. M. Choi, K. J. Shea, 2008J. Phy. Chem. C, 112, 18173.

27. J. P. Carpenter, C. M. Lukehart, S. R. Stock, J. E. Wittig, 1995Chem. Mater., 7, 201.

28. L. Pocard, D. C. Alsmeyer, R. L. Mc Creery, T. X. Neenan, M. R. Callstrom, 1992J. Am. Chem. Soc., 114, 769.

29. J. H. Kim, P. H. Holloway, 2005Adv. Mater., 17, 91.

30. H. R. Xia, G. W. Lu, P. Zhao, S. Q. Sun, X. L. Meng, X. F. Cheng, L. J. Qin, L. Zhu, Z. H. Yang, 2005J. Mater. Res., 20, 30.

31. S. Mukhopadhyay, K. P. Ramesh, R. Kannan, J. Ramakrishna, 2004Phys. Rev. B: Cond. Matter Mater. Phys., 70, 224202.

32. G. Dantelle, M. Mortier, D. Vivien, G. Patriarche, 2005J. Mater. Res., 20. 472.

33. S. Xiao, X. Yang, Z. Liu, X. H. Yan, 2004J. Appl. Phys., 96. 1360.

34. T. Buscaglia, V. Buscaglia, P. Ghigna, M. Viviani, G. Spinolo, A. Testino, P. Nanni, 2004Phys. Chem. Chem. Phys., 6, 3710.

35. S. Stepanov, E. Hernandez, M. Plata, 2004Opt. Lett., 29, 1327.

36. H. Lin, S. Jing, J. Wu, F. Song, N. Peyghambarian, E. Y. B. Pun, 2003J. Phys. D: Appl. Phys., 26, 812.

37. W. Sohler, H. Suche, 2000Opt. Eng., 66, 127.

38. W. Neuman, D. Pennington, J. Dawson, A. Drobshoff, R. Beach, I. Jovanovic, Z. Liao, S. Payne, C. P. J. Barty, 2005Proc. SPIE Int. Soc. Opt. Eng., 5653, 262.

39. X. Huang, Y. Liu, Z. Sui, M. Li, H. Lin, J. Wang, D. Zhao, F. Wang, J. Chen, 2005Proc. SPIE Int. Soc. Opt. Eng., 5623, 679.

40. C. Sanchez, B. Lebeau, F. Chaput, J. P. Boilot, 2003Adv. Mater., 15. 1969.

41. K. J. Shea, D. A. Loy, 2001Chem. Mater., 13, 3306.

42. H. Schmidt, 1989Sol-Gel Science and Technology: Published by World Scientific, Singapore.

43. Urquhart, 1988IEEE Proc.-J.; Optoelectronics, 135, 385.

44. C. B. Murray, C. R. Kagan, M. G. Bawendi, 1995Science, 270, 1335.

45. S. A. Empedocles, D. J. Norris, M. G. Bawendi, 1996Phys. Rev. Lett., 77, 3873.

46. L. Armelao, S. Gross, G. Obetti, E. Tondello, 2004Surf. Coat. Technol., 109, 218.

47. D. J. Di Giovanni, 1992Mater. Res. Soc. Proc., 244, 135.

48. R. Rossetti, R. Hull, J. M. Gibson, L. E. Brus, 1985J. Chem. Phys., 82 (l), 552; Zhang, Y., Raman, N., Bailey, J. K., Brinker, C. J., & Crooks, R. M. (1992) J. Phys. Chem., 96, 9098.

49. G. Schmid, 1992Chem. Rev., 92, 1709.

50. J. W. Grate, S. N. Kaganove, S. J. Patrash, R. Craig, M. Bliss, 1997Chem. Mater., 9 (5), 1201.

51. L. Ponson, N. Boechler, Y. M. Lai, M. A. Porter, P. G. Kevrekidis, C. Daraio, 2010Physical Review E, 82, 021301.

52. Spadoni, C. Daraio, 2010Proc. Natl. Acad. Sci. USA, 107, 7230.

53. M. A. Porter, C. Daraio, E. B. Herbold, I. Szelengowicz, P. G. Kevrekidis, 2008Physical Review E, 77, 015601(R).

54.   A. Sullivan, B. C. Olbricht, L. A. Dalton, 2008Journal of Lightwave Technology, 26 (15), 2345.

55.   H. , B. Chen, S. Takafumi, L. R. Dalton, A. K. Jen, Y. , 2001J. Am. Chem. Soc., 123, 986.

56. $\eta 6$ -Bisarene chromium complexes decompose in dilute acid.Oxidation and disproportion products (of both organic and Cr) are observed. [57] η6-Arene Cr(CO)3 complexes form adducts with Lewis and protonic acids but their decomposition products have not been investigated. [58]

57.   B. G. Gribov, D. D. Mozzhukhin, B. I. Kozyrkin, A. S. Strizhkova, 1972J. Gen. Chem. USSR (Eng. Transl.), 42, 2521.

58.   T. C. Flood, E. Rosenberg, A. Sarhangi, 1997J. Am. Chem. Soc., 99, 4334; Lillya, C. P. & Sahatjian, R. A. (1972) Inorg. Chem., 11, 889.

Chapter 4

ATOMIC AND MOLECULAR LOW-N RYDBERG STATES IN NEAR CRITICAL POINT FLUIDS

Luxi Li, Xianbo Shi[1], Cherice M. Evans[2] and Gary L. Findley[3]

[1]Brookhaven National Laboratory, Upton, NY, USA

[2]Department of Chemistry, Queens College – CUNY and the Graduate Center – CUNY, New York, NY, USA

[3]Chemistry Department, University of Louisiana at Monroe, Monroe, LA, USA

INTRODUCTION

Since electronic excited states are sensitive to the local fluid environment, dopant electronic transitions are an appropriate probe to study the structure of near critical point fluids (i.e., perturbers). In comparison to valence states, Rydberg states are more sensitive to their environment [1]. However, high-n Rydberg states are usually too sensitive to perturber density fluctuations, which makes these individual dopant states impossible to investigate. (Nevertheless, under the assumption that high-n Rydberg state energies behave similarly to the ionization threshold of the dopant, dopant high-n Rydberg state behavior in supercritical fluids can be probed indirectly by studying the energy of the quasi-free electron, through photoinjection [2–11] and field ionization [12–19].) Low-n Rydberg states, on the other hand, are excellent spectroscopic probes to investigate excited state/fluid interactions. The study of low-n Rydberg states in dense fluids began with the photoabsorption of alkali metals in rare gas fluids [20, 21]. Later researchers expanded the investigation into rare gas dopants in supercritical rare gas fluids [22–26], and molecular dopants in atomic and molecular perturbers [27–36]. However, none of these previous groups studied dilute solutions near the critical point of the solvent. (The single theoretical study of a low-n Rydberg state in a near critical point fluid was performed by Larrégaray, et al. [35]; this investigation predicted a change in the line shape and in the perturber induced shift of the Rydberg transition.)

These results from previous experimental and theoretical investigations of low-n Rydberg states in dense fluids are reviewed in Section 2. In this Chapter, we present a systematic investigation of the photoabsorption of atomic and molecular dopant low-n Rydberg transitions in near critical point atomic fluids [37–40]. The individual systems probed allowed us to study (dopant/ perturber) atomic/atomic interactions (i.e., Xe/Ar) and molecular/atomic inter- actions (i.e., CH_3I/Ar, CH_3I/Kr, CH_3I/Xe) near the perturber critical point. The experimental techniques and theoretical

methodology for this extended study of dopant/perturber interactions are discussed in Section 3. Section 4 presents a review of our results for low-n atomic and molecular Rydberg states in atomic supercritical fluids. The accuracy of a semi-classical statistical line shape analysis is demonstrated, and the results are then used to obtain the perturber-induced energy shifts of the primary low-n Rydberg transitions. A striking critical point effect in this energy shift was observed for all of the dopant/perturber systems presented here. A discussion of the ways in which the dopant/perturber interactions influence the perturber-induced energy shift is also presented in Section 4. We conclude with an explanation of the importance of the inclusion of three-body interactions in the line-shape analysis, and with a discussion of how this model changes when confronted with non-spherical perturbers and polar fluids.

PERTURBER EFFECTS ON LOW-N RYDBERG STATES

2.1. Supercritical fluids A supercritical fluid (SCF) exists at a temperature higher than the critical temperature and, therefore, has properties of both a liquid and a gas [41]. For example, SCFs have the large compressibility characteristic of gases, but the potential to solvate materials like a liquid. Moreover, near the critical point the correlation length of perturber molecules becomes unbounded, which induces an increase in local fluid inhomogeneities [41]. The local density $\rho(r)$ of a perturbing fluid around a central species (either the dopant or a single perturber) is defined by [42, 43]

$$\rho(r) = g(r)\,\rho\,,$$

where g(r) is the radial distribution function and ρ is the bulk density. For neat fluids and for most dilute dopant/perturber systems, the interactions between the species in the system are attractive in nature. Thus, the perturber forms at least one solvent shell around the dopant (or a central perturber). As the dilute dopant/perturber system approaches the critical density and temperature of the perturber, the intermolecular interactions increase. This increase leads to a change in the behavior of the local density as a function of the bulk density [41]. Therefore, the changes in fluid properties near the critical point are due to the higher correlation between the species in the fluid.

These intermolecular correlations are usually probed spectroscopically by dissolving a dopant molecule in the SCF. Since electronic excited states are incredibly sensitive to local fluid environment, fluctuations in the fluid environment can be investigated by monitoring variations in the absorption or emission of the dopant. Fig. 1 gives three example spectroscopic studies of a dopant in near critical point SCFs. Fig 1a shows the energy position of an anthracene emission line for anthracene doped into near critical point carbon dioxide [44]. Unfortunately, no emission data on non-critical isotherms were measured in this fluorescence emission study. However, the maximum emission position shows a striking critical effect in comparison to a calculated baseline. A more complete investigation of temperature effects on the local density of SCFs via UV photoabsorption [45] of ethyl p-(N,N-dimethylamino) benzoate (DMAEB) in supercritical CHF_3 is presented in Fig. 1b for three isotherms near the critical isotherm. The isotherms shown are at the reduced temperature T_r [where $T_r \equiv T/T_c$ with T_c being the critical temperature of the fluid] of 1.01, 1.06 and 1.11. Although the three isotherms are evenly spaced, the photoabsorption shifts are very similar on the $T_r = 1.06$ and $T_r = 1.11$ isotherms, while the shift on the $T_r = 1.01$ isotherm is significantly different. Therefore, the perturber induced shift is only temperature sensitive near the critical point of the perturber. Urdahl, et al. [46, 47] studied the $W(CO)_6$ T_{1u} asymmetric CO stretching mode doped into supercritical carbon dioxide, ethane and trifluoromethane by IR photoabsorption. All three systems show similar behavior, an example of which is presented in Fig. 1c.

Again, the absorption band position changes significantly along the critical isotherm near the critical density. However, extracting the data presented in Fig. 1 is difficult. Moreover, most of the previous work [41] has used dopant valence transitions as probes, and these states tend to be less sensitive to local environments [1]. In the present work, we investigate near critical point SCFs using low-n Rydberg state transitions as the probe. Since Rydberg states are hydrogen-like, it should be possible to model these states within the statistical mechanical theory of spectral line-broadening.

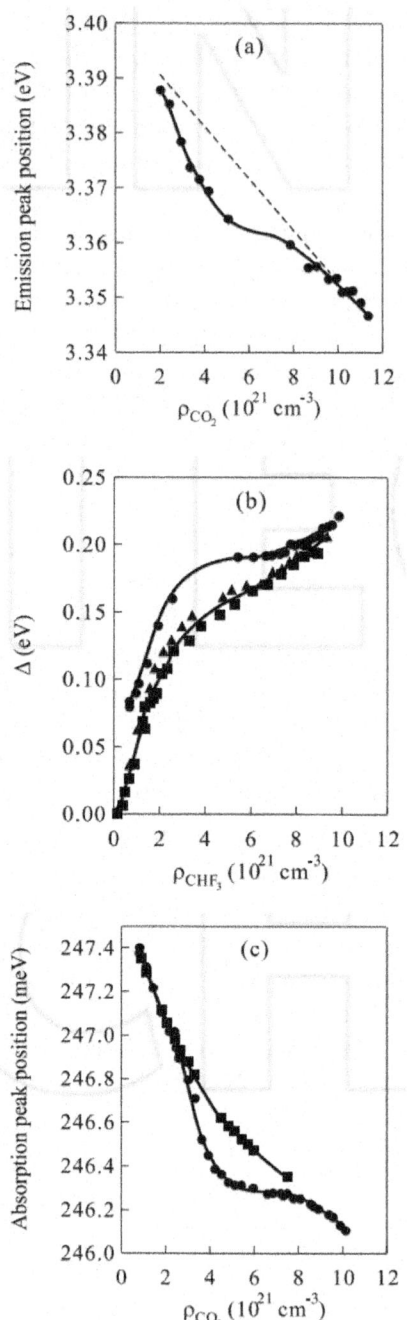

Figure 1: (a) The energy of fluorescence emission for (•) anthracene doped into supercritical carbon dioxide at a reduced temperature T_r 1.01 plotted as a function of carbon

dioxide number density ρCO_2 . The dashed line (- - -) is a reference line calculated using perturber bulk densities. Adapted from [44]. (b) The perturber induced energy shifts of the photoabsorption maximum for DMAEB doped into supercritical CHF3 at (•) 30.0∘C [i.e., T_r 1.01], () 44.7∘C, and () 59.6∘C, plotted as a function of CHF_3 number density ρCHF_3 . Adapted from [45]. (c) The energy position of infrared absorption for the $W(CO)_6$ CO stretching mode in supercritical carbon dioxide at (•) 33∘C [i.e., T_r 1.01] and at () 50∘C, plotted as a function of ρCO_2 . Adapted from [46, 47]. The solid lines in (a) - (c) are provided as a visual aid. 2.2.

Theoretical model

Due to the hydrogenic properties of Rydberg states, the optical electron is in general insensitive to the structure of the cationic core. Therefore, both atomic and molecular Rydberg transitions can be modeled within the same theory. In a very dilute dopant/perturber system, assuming that the dopant Rydberg transition is at high energy [i.e. $\beta (E_g - E_e) \gg 1$, , $\beta = 1/(kT)$], the Schrödinger equation is

$$H|\Psi\rangle = E|\Psi\rangle ,$$

(1)

where the eigenfunction $|\Psi\rangle$ is a product of the dopant electronic wavefunction $|\alpha\rangle$ and the individual perturber wavefunctions $|\psi i; \alpha\rangle$. The Hamiltonian H in eq. (1) is the sum of several individual Hamiltonians, namely the Hamiltonian for the free dopant, the Hamiltonian for the free perturber, and the Hamiltonian for the dopant/perturber intermolecular correlation. The Hamiltonian HFD for the free dopant is [48]

$$H_{FD} = \sum_\alpha E_\alpha |\alpha\rangle\langle\alpha| ,$$

(2)

where α = e, g with e and g representing the excited state and ground state of the free dopant. The Hamiltonian HFP for the free non-interacting perturber is given by [48]

$$H_{FP} = - \sum_i^N \frac{\hbar^2}{2m} \nabla_i^2 ,$$

(3)

where N is the total number of perturbing atoms in the range of the Rydberg optical electron. Finally, the Hamiltonian H_{PD} for the intermolecular interaction is [48]

$$H_{PD} = \sum_\alpha \sum_i [V_\alpha(r_i) + V'(r_i)] |\Psi\rangle\langle\Psi| ,$$

(4)

where V(r) is the dopant/perturber intermolecular potential, and V' (r) is the perturber/perturber intermolecular potential. Therefore, the total Hamiltonian H is [48]

$$H = H_{FD} + H_{FP} + H_{PD},$$

(5)

which can be rewritten as a ground state Hamiltonian expectation value H_g and an excited state Hamiltonian expectation value H_e. The absorption coefficient function is given as the Fourier transform [48]

$$\mathfrak{L}(\omega) \equiv \frac{1}{2\pi} \int_{-\infty}^{\infty} dt\, e^{-i\omega t} \langle\, \overrightarrow{\mu}(0) \cdot \overrightarrow{\mu}(t)\, \rangle,$$

(6)

where the autocorrelation function (i.e $\langle \cdots \rangle$) is the thermal average of the scalar product of the dipole moment operator (i.e., $\overrightarrow{\mu}$) of the dopant at two different times. This autocorrelation function can be resolved within the Liouville operator formalism to give [48]

$$\langle\, \overrightarrow{\mu}(0) \cdot \overrightarrow{\mu}(t)\, \rangle \equiv \exp\left[\rho_P \langle e^{i L_g t} - 1\rangle_g\right],$$

(7)

where the two-body Liouville operator L_g is defined by

$$L_g \Omega = H_e \Omega - \Omega H_g = [H_g, \Omega] + (E_e - E_g)\Omega,$$

(8)

where Ω is an arbitrary operator. However, if lifetime broadening is neglected, only the dopant/perturber interaction and the dopant electronic energy change during the transition. Therefore, the autocorrelation function can be rewritten as [48]

$$\langle\, \overrightarrow{\mu}(0) \cdot \overrightarrow{\mu}(t)\, \rangle \equiv e^{i\omega_0 t} \exp\left[\rho_P \langle e^{i\Delta V t} - 1\rangle_g\right],$$

(9)

where ω_0 is the transition frequency of the neat dopant, ρ_P is the perturber density, and $\Delta V = V_e - V_g$, with V_e and V_g, being the excited-state dopant/ground-state perturber and ground-state dopant/ground-state perturber intermolecular potentials, respectively. In semi-classical line shape theory, the line shape function for an allowed transition is given by [20],

$$\mathfrak{L}(\omega) = \frac{1}{2\pi} \int_{-\infty}^{\infty} dt\, e^{-i[\omega(\mathbf{R}) - \omega_0]t}\, \frac{Z(\beta V_g + it\Delta V)}{Z(\beta V_g)},$$

(10)

where Z is the partition function and R denotes the collection of all dopant/perturber distances. Under the classical fluid approximation of Percus [49–51], the autocorrelation function $\Phi(t)$ is given by a density expansion [20]

$$\Phi(t) = \ln Z(\beta V_g + it\Delta V) - \ln Z(\beta V_g) = A_1(t) + A_2(t) + \cdots,$$

(11)

where [20, 25]

$$A_n(t) = \frac{1}{n!} \int \cdots \int \prod_{j=1}^{n} d^3 R_j \, \mathfrak{F}(\mathbf{R}_1, \ldots, \mathbf{R}_n)$$

$$\times \prod_{j=1}^{n} \left[\exp(-i \, \Delta V(\mathbf{R}_j) \, t) - 1 \right] . \tag{12}$$

In eq. (12), $\mathfrak{F}(\mathbf{R}_1, \ldots, \mathbf{R}_n)$ is the Ursell distribution function [25, 26, 49–51], and $\Delta V(R) = V_e(\mathbf{R}) - V_g(\mathbf{R})$. The Ursell distribution function for two body interactions [25, 26] is $\mathfrak{F}(\mathbf{R}) = \rho_P \, g_{PD}(\mathbf{R})$, where $g_{PD}(\mathbf{R})$ is the perturber/ dopant radial distribution function. The three body Ursell distribution function is estimated using the Kirkwood superposition approximation [52] to be

$$\mathfrak{F}(\mathbf{R}_1, \mathbf{R}_2) = \rho_P^2 \, [\, g_{PP}(|\mathbf{R}_1 - \mathbf{R}_2|) - 1 \,] \, g_{PD}(\mathbf{R}_1) \, g_{PD}(\mathbf{R}_2) \,,$$

(13)where $g_{PP}(R)$ is the perturber/perturber radial distribution function. The density expansion terms are multibody interactions between dopant and perturber over all space. $A_1(t)$ is the dopant/perturber two-body interaction, $A_2(t)$ is the dopant/perturber three-body interaction, and $A_n(t)$ is the dopant/perturber $n + 1$ body interaction. Substitution of these Ursell distribution functions into eq. (12) under the assumption of spherically symmetric potentials, gives [25, 26, 37]

$$A_1(t) = 4 \pi \rho_P \int_0^\infty dr \, r^2 \, g_{PD}(r) \left[e^{-it \Delta V(r)} - 1 \right] , \tag{14}$$

And

$$A_2(t) = 4 \pi \rho_P^2 \int_0^\infty dr_1 \, r_1^2 \, g_{PD}(r_1) \left[e^{-it \Delta V(r_1)} - 1 \right]$$

$$\times \int_0^\infty dr_2 \, r_2^2 \, g_{PD}(r_2) \left[e^{-it \Delta V(r_2)} - 1 \right]$$

$$\times \frac{1}{r_1 \, r_2} \int_{|r_1 - r_2|}^{|r_1 + r_2|} s \, [\, g_{PP}(s) - 1 \,] \, ds . \tag{15}$$

Since the strength of the interaction decreases as the number of bodies involved increases, and since higher order interactions are more difficult to model, most line shape simulations are truncated at the second term $A_2(t)$. The autocorrelation function can be written as a power series expansion at $t = 0$, namely [21, 48, 53, 54]

$$\Phi(t) = \sum_{n=1}^{\infty} \frac{i^n}{n!} \, m_n \, t^n , \tag{16}$$

where the expansion coefficients m_i are given by

$$m_n = \frac{1}{(\sqrt{-1})^n} \frac{d^n}{dt^n} \Phi(t) \bigg|_{t=0} .$$

(17)

Using eq. (11) with $A_1(t)$ and $A_2(t)$ from eqs. (14) and (15), the first two expansion coefficients become [25]

$$m_1 = -4\pi\rho_P \int_0^\infty dr\, r^2\, g_{PD}(r)\, \Delta V(r) ,$$

(18)

And

$$m_2 = m_1^2 + 4\pi\rho_P \int_0^\infty dr\, r^2\, g_{PD}(r)\, \Delta V(r)^2$$

$$+ 8\pi^2 \rho_P^2 \int_0^\infty \int_0^\infty dr_1\, dr_2\, r_1\, r_2\, g_{PD}(r_1)\, g_{PD}(r_2)$$

$$\times\ \Delta V(r_1)\, \Delta V(r_2) \int_{|r_1-r_2|}^{|r_1+r_2|} s\, [\, g_{PP}(s) - 1\,]\, ds .$$

(19)

The above expansion coefficients are equivalent to the moments of the optical coefficient, which are defined as [25]

$$m_n = \int \mathcal{L}(E)\, E^n\, dE .$$

(20)

Thus, the perturber induced shift $\Delta(\rho_p)$ in the energy position of the optical coefficient maximum is [25, 26]

$$\Delta(\rho_P) = M_1 = \frac{m_1}{m_0} = \int \mathcal{L}(E)\, E\, dE \bigg/ \int \mathcal{L}(E)\, dE ,$$

(21)

while the full-width-half-maximum of the experimental absorption spectrum is proportional to

$$M_2 = \left[\frac{m_2}{m_0} - M_1^2\right]^{1/2} = \left[\left(\int \mathcal{L}(E)\, E^2\, dE \bigg/ \int \mathcal{L}(E)\, dE\right) - M_1^2\right]^{1/2} ,$$

(22)

where $\mathcal{L}(E)$ is the absorption band for the transition and $E = \hbar(\omega - \omega_0)$ [$\hbar \equiv$ reduced Planck constant]

PREVIOUS STUDIES

The interaction of dopant low-n Rydberg states with dense fluids has previously been the subject of some interest both experimentally and theoretically [1, 22–36]. The first detailed investigation of these interactions was a study of

X_e low-n Rydberg states doped into the dense rare gas fluids (i.e., argon, neon and helium) by Messing, et al. [25, 26]. In the same year, Messing, et al. [27, 28] presented their studies of molecular low-n Rydberg states in dense A_r and Kr. A decade later, Morikawa, et al. [34] probed the NO valence and low-n Rydberg state transitions in dense argon and krypton. All of these experiments [25–28, 34] used a basic moment analysis of absorption bands to determine the energy shifts of these bands as a function of perturber number density. The photoabsorption energy shifts were then simulated [25–28, 34] using eq. (18) for various assumptions of intermolecular potentials and radial distribution functions. Messing, et al. also performed line shape simulations under the assumption of a Gaussian line shape [25–28] for selected perturber number densities. As molecular dynamics developed, research groups [30, 31, 33, 35, 36] returned to absorption line shape simulations in an attempt to match the asymmetric broadening observed experimentally.

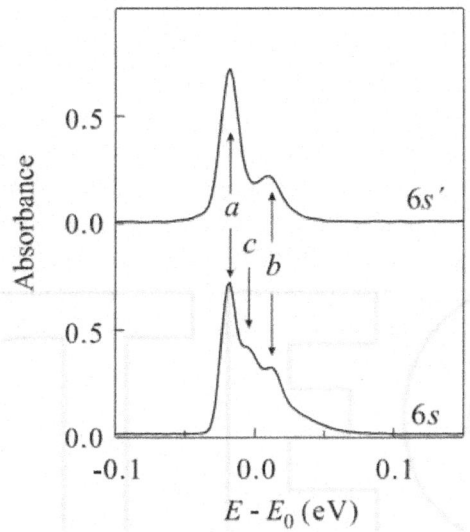

Figure 2: Photoabsorption spectra (relative units) of the Xe 6s and 6's transitions doped into A_r at an argon number density of $\rho A_r = 1.47 \times 1021 \text{ cm}^{-3}$ and a temperature of 23.6°C. a corresponds to eq. (23), b to eq. (24), and c to eq. (25). For the 6s Rydberg state $E_0 = 8.437$ eV and for the 6's Rydberg state $E_0 = 9.570$ eV. Data from the present work.

When Xe 6's and 6s Rydberg state transitions are excited in the presence of low density argon or krypton, satellite bands appear on the higher energy wing of the absorption or emission spectra [22–24]. These blue satellite bands increase with increasing perturber density [22– 24] and with decreasing temperature. Therefore, Castex, et al. [22–24] concluded that these blue satellites are caused by the formation of dopant/perturber ground state and excited state dimers. An example of the X_e 6's and 6s Rydberg transitions in the presence of argon is shown in Fig. 2. The primary X_e absorption transition, or [37]

$$Xe + h\nu \xrightarrow{Ar} Xe^*,$$

(23)

is indicated in Fig. 2 as a. The X_eAr dimer transitions that yield the blue satellite bands are [22–24]

$$XeAr + h\nu \xrightarrow{Ar} Xe^* + Ar,$$

(24)

And

$$XeAr + h\nu \xrightarrow{Ar} Xe^*Ar.$$

(25)

These transitions are indicated in Fig. 2 as b and c, respectively. Detailed investigations [25] of the Xe '6s Rydberg state doped in supercritical argon indicated that the energy position at the photoabsorption peak maximum shifted slightly to the red at low argon density and then strongly to the blue (cf. Fig. 3a). Similar results, which are shown in Fig. 3b, were then observed for the Xe 6's Rydberg states in supercritical argon [26]. These studies [25, 26] concluded that both the perturber-induced energy shift and the line shape broadening were temperature independent. However, since the blue satellite bands grow and broaden as a function of perturber number density, the energy position of the maximum absorption (or the first moment from a moment analysis) is not an accurate energy position for the primary Xe Rydberg transition. Thus, modeling the experimental first moment M_1 and second moment M_2 using eqs. (19) and (20) required three groups of intermolecular potential parameters for different perturber density ranges.

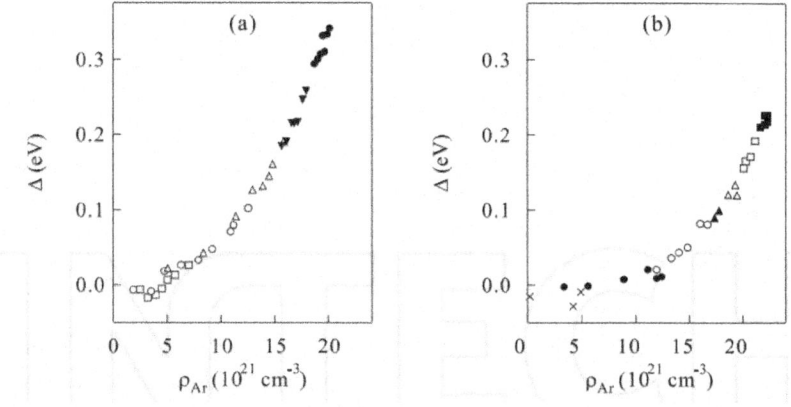

Figure 3: The perturber induced shift Δ of (a) the Xe 6s [25] and (b) the Xe 6s [26] absorption maximum plotted as a function of argon num-

ber density ρAr at different temperatures. In (a), the markers are (□) 25°C; (○) −83°C; (△) −118°C; (▼) −138°C; and (●) −163°C. In (b), the markers are for temperature ranges of (×) −23°C to 27°C; (●) −93°C to −63°C; (○) −123°C to −113°C; (▲) −138°C to −128°C; (△) −158°C to −148°C; (□) −173°C to −163°C; and (■) −186°C to −178°C.

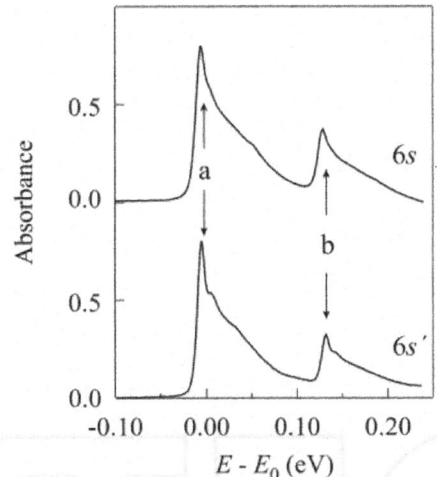

Figure 4: Photoabsorption spectra (relative units) of the CH3I 6s and 6s transitions doped into A_r at an argon number density of $\rho A_r = 1.89 \times 10^{21}$ cm−3 and a temperature of −79.8∘C. a corresponds to eq. (26) and b to eq. (27). For the 6s Rydberg state E_0 = 6.154 eV and E_0 = 6.767 eV for the 6s Rydberg state. Data from the present work.

The Xe 6s and 6's Rydberg transitions [26] in supercritical helium and neon show a similar perturber density dependence as that shown in Fig. 3. However, these two systems do not form ground state or excited state dimers and, therefore, do not have blue satellite bands. Thus, the moment analysis of the photoabsorption spectra presented a more accurate perturber induced shift as a function of perturber number density. Because of the

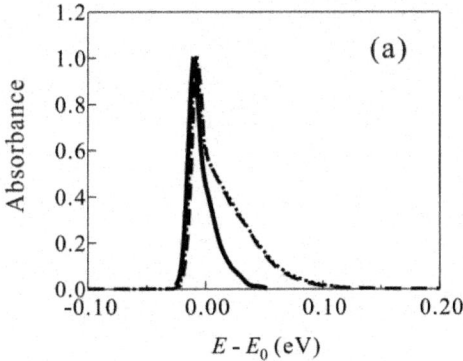

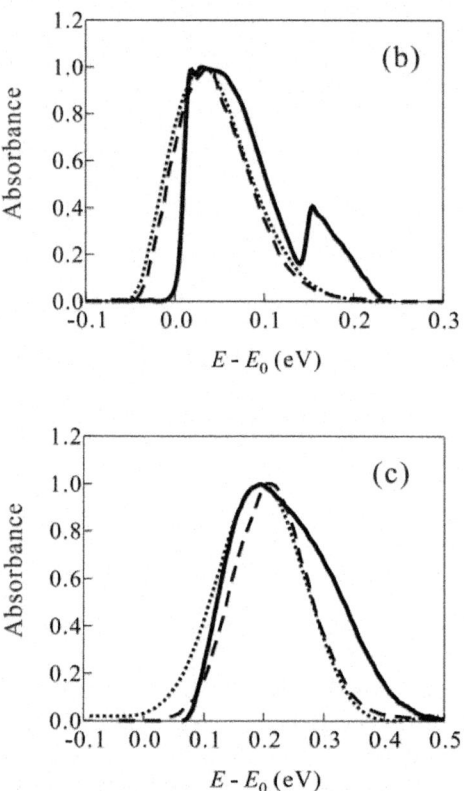

Figure 5: The line shape simulation of the CH_3I 6s transition doped into supercritical argon using (- - -) a semi-classical line shape function data [33] and using (···) molecular dynamics [32] in comparison to (—) the photoabsorption spectra at (a) $\rho A_r = 2.0 \times 1021$ cm^{-3}, (b) $\rho A_r = 7.6 \times 1021$ cm^{-3}, and (c) $\rho A_r = 2.0 \times 1022$ cm^{-3}. Experimental data are from the present work.

simplicity of the absorption bands, Messing, et al. [26] simulated the line shapes of the Xe 6s and 6's transitions in both helium and neon at a single perturber number density using eqs. (10) and (11). These simulations indicated that an accurate line shape could be obtained without blue satellite bands [26]. Unfortunately, no temperature dependence was reported in these papers [25, 26].

CH_3I in Ar and Kr

Since CH_3I is a molecular dopant, vibrational transitions as well as the adiabatic transition appear in the photoabsorption spectra. The adiabatic transition is given by [37]

$$CH_3I + h\nu \longrightarrow CH_3I^*,$$ (26)

denoted a in Fig. 4, as well as one quantum of the CH_3 deformation vibrational transition v2 in the excited state, or

$$CH_3I + h\nu \longrightarrow CH_3I^* (\nu_2),$$ (27)

denoted b, in Fig. 4.

Although vibrational transitions are apparent, CH_3I [27–29, 32] has been investigated extensively because of the "atomic" like nature of the adiabatic 6s and 6's Rydberg transitions. Messing, et al. [27, 28] extracted the perturber dependent shift $\Delta(\rho_p)$ of the CH_3I 6s and 6s Rydberg states by performing a moment analysis on the photoabsorption bands using eq. (21). This analysis indicated that $\Delta(\rho_p)$ tended first to lower energy and then to higher energy as ρ_p increased from low density to the density of the triple-point liquid, similar to the trends shown in Fig. 3 for X_e in Ar. However, Messing, et al. [27, 28] did not explore critical temperature effects on $\Delta(\rho_p)$, nor did they correctly account for the vibrational bands on the blue side of the adiabatic Rydberg transition. Messing, et al. [28] did attempt to model the CH_3I 6s Rydberg transition in argon using a semi-classical statistical line shape function under the assumption that the adiabatic and vibrational transitions have exactly the same line shapes, although no comparison between the experimental spectra and the simulated line shapes was provided. Later researchers [30–33] concentrated on the simulation of the CH_3I 6s Rydberg state doped into argon using both molecular dynamics and semi-classical integral methods. Egorov, et al. [33] showed that the semi-classical integral method can yield results comparable to the molecular dynamics calculations of Ziegler, et al. [30–32]. Comparing the semi-classical method and molecular dynamics simulation to experimental spectra (cf. Fig. 5) shows that

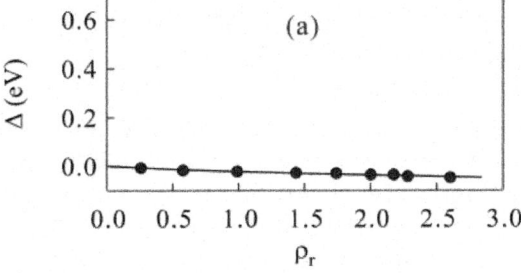

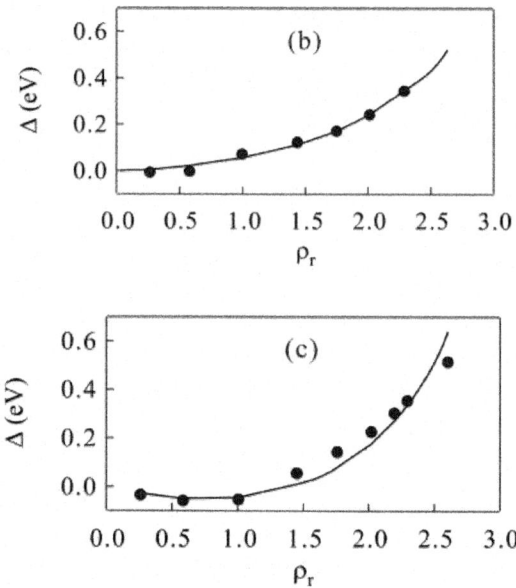

Figure 6: Experimental (•) and calculated (—) energy shift of the NO (a) $B'^2\Delta \leftarrow X^2\Pi \, (v' = 7, 0)$ valence transition, (b) $A^2\Sigma^+ \leftarrow X^2\Pi \, (v' = 0, 0)$ Rydberg transition and (c) $C^2\Pi^+ \leftarrow X^2\Pi \, (v' = 0, 0)$]

Rydberg transition plotted versus the reduced argon number density [34]. both methods can be used to simulate the experimental spectra with an appropriate choice of intermolecular potentials.

No In Ar

Morikawa, et al. [34] investigated valence and low-n Rydberg transitions doped into supercritical argon. They used eq. (21) to determine the perturber induced shift $\Delta(\rho_p)$ for several low-n Rydberg transitions as well as a valence transition. Under the assumption of spherically symmetric potentials for the ground and excited states of the NO/A$_r$ systems, an accurate moment analysis using eq. (18) was performed (cf. Fig. 6). In The NO valence state transition (cf. Fig. 6a) shows a slight perturber-induced red shift, which differs significantly from the obvious blue shift of the low-n NO Rydberg state transitions shown in Figs. 6b and c. Fig. 6 also shows that low-n Rydberg states make a more sensitive probe to perturber effects than valence transitions. Later groups [35, 36] did line shape simulations to model the experimental spectra. Larrégaray, et al. [35] measured the NO 3sσ transition doped into supercritical argon at selected argon densities. Then, using molecular dynamics they successfully modeled the line shape of the transition. Once the intermolecular potentials and boundary conditions for the molecular dynamics simulations had been set against experimental data, Larrégaray, et al. [35] calculated the line shape for NO in A_r at the critical density and temperature. These calculations predicted that the photoabsorption peak maximum position would shift to the blue when argon was near its critical point. Later, Egorov, et al. [36] showed that similar results could be obtained using the semi-classical approximation. Therefore, line shape simulations using molecular dynamics and semi-classical line shape theory show comparable results.

EXPERIMENT TECHNIQUES AND THEORETICAL METHODOLOGY

Experimental techniques

All of the photoabsorption measurements presented in Sections 4 and 5 were obtained using the one-meter aluminum Seya-Namioka (Al-SEYA) beamline on bending magnet 8 of the Aladdin storage ring at the University of Wisconsin Synchrotron Radiation Center (SRC) in Stoughton, WI. This beamline, which was decommissioned during Winter 2007, produced monochromatic synchrotron radiation having a resolution of ~ 8 meV in the energy range of 6 - 11 eV. The monochromatic synchrotron radiation enters the vacuum chamber which is equipped with a sample cell (cf. Fig. 7). The photoabsorption signal is detected by a photomultiplier that is connected to the data collection computer via a Keithley 486 picoammeter. The pressure in the vacuum chamber is maintained at low 10^{-8} to high 10^{-9}.

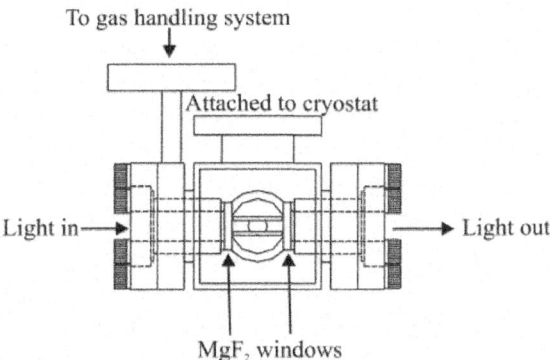

Figure 7: Schematic of the copper experimental cell.

Torr by a Perkin-Elmer ion pump. The experimental copper cell is equipped with entrance and exit MgF_2 windows (with an energy cut-off of 10.9 eV) that are capable of withstanding pressures of up to 100 bar and temperatures of up to 85∘C. This cell, which has a path length of 1.0 cm, is connected to an open flow liquid nitrogen cryostat and resistive heater system allowing the temperature to be controlled to within ± 0.5∘C with a Lakeshore 330 Autotuning Temperature controller. The gas sample is added through a gas handling system (GHS), which consists of 316-stainless steel components connected by copper gasket sealed flanges. The initial pressure for the GHS and sample cell is in the low 10−8 Torr range.

During the initial bake out, the GHS and the vacuum chamber were heated to 100∘C under vacuum for several days to remove any water adsorbed onto the surface of the stainless steel. The initial bake was stopped when the base pressure of the GHS and sample chamber is in the low 10^{-7} or high 10^{-8} Torr range, so that upon cooling the final GHS base pressure was $10^{-8} - 10^{-9}$ Torr. Anytime a system was changed (either the dopant or the perturber), the GHS was again baked in order to return the system to near the starting base pressure. This prevented cross-contamination between dopant/perturber systems.

The intensity of the synchrotron radiation exiting the monochromator was monitored by recording the beam current of the storage ring as well as the photoemission from a nickel mesh situated prior to the sample cell. The light then entered the experimental cell through a MgF_2 window, traveled through the sample and then a second MgF_2 window (cf. Fig. 7) before striking a thin layer of sodium salicylate powder on the inside of a glass window that preceded the photomultiplier tube. An empty cell (acquired for a base pressure of 10^{-7} or 10^{-8} Torr) spectrum was used to correct the dopant absorption spectra for the monochromator flux, for absorption by the MgF_2 windows, and for any fluctuations in the quantum efficiency of the sodium salicylate window.

Two dopants (i.e. methyl iodide and xenon) and five perturbers (i.e. argon, krypton, xenon, carbon tetrafluoride and methane) were investigated. All dopants and perturbers were used without further purification: methyl iodide (Aldrich, 99.45%), argon (Matheson Gas Products, 99.9999%), krypton (Matheson Gas Products, 99.998%), and xenon (Matheson Gas Products, 99.995%). When CH_3I was the dopant, the CH_3I was degassed with three freeze/pump/thaw cycles prior to use. Photoabsorption spectra for each neat dopant and each neat perturber were measured to verify the absence of impurities in the spectral range of interest. The atomic perturber number densities were calculated from the Strobridge equation of state [55] with the parameters obtained from [56] for argon, [57] for krypton, and [58] for xenon. All densities and temperatures were selected to maintain a single phase system in the sample cell.

At temperatures below T_c, a change in density required a change in temperature, since the isotherms are steeply sloped. The sensitivity of the absorption spectra to local density required that the quality of a data set be monitored by performing basic data analysis during measurements. Any anomalies were corrected by immediately re-measuring the photoabsorption spectrum for the problem density/temperature/pressure after allowing additional time for increased density stabilization. Once a data set was obtained for non-critical temperatures, the photoabsorption data for perturber densities on an isotherm near the critical isotherm were then measured. For the near critical data set we selected a temperature that was +0.5°C above the critical temperature (chosen to prevent liquid formation in the cell during temperature stabilization near the critical density and to minimize critical opalescence during data acquisition). Near the critical density, the consistency of the density step is dependent on the slope of the critical isotherm. If the critical isotherm has a small slope in this region, it becomes difficult to acquire samples at a constant density step size due to our inability to vary the perturber pressure practically by less than 0.01 bar and to the difficulties encountered in maintaining temperature stability. For instance, near the critical density of xenon, a 1 mbar change in pressure or a 0.001°C change in temperature causes a density change of 2.0×10^{21} cm^{-3}. Maintaining the necessary temperature stability (i.e., ± 0.2°C) during the acquisition of data along the critical isotherm is difficult with an open flow liquid nitrogen cryostat system and usually required constant monitoring with manual adjustment of the nitrogen flow.

THEORETICAL METHODOLOGY

Line shape function

The experimental line shapes were simulated using the semi-classical statistical line shape function given in eq. (10). Rewriting eq. (10) in terms of the autocorrelation function allows eq. (10) to be given as a Fourier transform, namely [25, 26, 33, 37]

$$\mathfrak{L}(\omega) = \frac{1}{2\pi} \, \text{Re} \int_{-\infty}^{\infty} dt \, e^{-i\omega t} \langle e^{i\omega(R)t} \rangle \, ,$$

$$(28)$$

where $\omega = \omega(R) - \omega_0$, with ω_0 being the transition frequency for the neat dopant. Eq. (28) neglects lifetime broadening and assumes that the transition dipole moment is independent of R. In the substitution of the exponential density expansion [i.e., eq. (11)] for the autocorrelation function, the general term A_n represents a (n+1)-body interaction [20]. However, since the strength of the interaction decreases as the number of bodies involved increases, and since the higher order interactions are more difficult to model, our line shape simulations are truncated at the second term $A_2(t)$, or three body interactions. Within this approximation, eq. (28) becomes

$$\mathcal{L}(\omega) = \frac{1}{2\pi} \text{Re} \int_{-\infty}^{\infty} dt\, e^{-i\omega t}\, \exp\left[A_1(t) + A_2(t)\right]$$

(29)

where the two terms are recalled from eq. (14),

$$A_1(t) = 4\pi\rho_P \int_0^{\infty} dr\, r^2\, g_{PD}(r) \left[e^{-it\Delta V(r)} - 1 \right] \quad,$$

and eq. (15),

$$A_2(t) = 4\pi\rho_P^2 \int_0^{\infty} dr_1\, r_1^2\, g_{PD}(r_1) \left[e^{-it\Delta V(r_1)} - 1 \right]$$
$$\times \int_0^{\infty} dr_2\, r_2^2\, g_{PD}(r_2) \left[e^{-it\Delta V(r_2)} - 1 \right]$$
$$\times \frac{1}{r_1 r_2} \int_{|r_1 - r_2|}^{|r_1 + r_2|} s\left[\, g_{PP}(s) - 1 \,\right] ds \,.$$

The required radial distribution functions (i.e., g_{PP} and g_{PD}) were obtained from the analytical solution of the Ornstein-Zernike equation for a binary system within the Percus-Yevick (PY) closure [59], while the Fourier transform for eq. (29) was performed using a standard fast Fourier transform algorithm [60]. The line shape obtained from the transform of eq. (29) was convoluted with a standard Gaussian slit function to account for the finite resolution (~ 8 meV) of the monochromator. More detailed discussions are given below.

Fast Fourier transform

A Fourier transform has the general form [60]

$$F(\omega) = \frac{1}{2\pi} \int_{-\infty}^{\infty} f(t)\, e^{-i\omega t}\, dt \,.$$

(30)

Since the line shape function is calculated numerically, the integration limits for eq. (30) must be finite and, therefore, an appropriate integration range must be determined. For any Fourier transformation, the integration limit and the total number of steps are related through [60]

$$\delta t \times \delta \omega = \frac{2\pi}{N} \,,$$

(31)

where δ stands for the sampling interval (i.e., the step size) of the corresponding variable and N is the total number of discrete points. Fourier transforms rely on the fact that data are usually obtained in discrete steps and the generating functions f(t) and F(ω) can be represented by the set of points

$$f_k \equiv f(t_k), \quad t_k = k\,\delta t, \quad k = 1, \ldots, N,$$

$$F_n \equiv F(\omega_n), \quad \omega_n = n\,\delta\omega, \quad n = -\frac{N}{2}, \ldots, \frac{N}{2} - 1. \tag{32}$$

Therefore, the function F(ω) is determined point-wise using

$$F(\omega_n) = \frac{1}{2\pi} \int_{-\infty}^{\infty} f(t)\,e^{-i\omega_n t}\,dt = \frac{1}{2\pi} \sum_{k=1}^{N} f_k\,e^{-i\omega_n t_k}\,\delta t$$

$$= \frac{\delta t}{2\pi} \sum_{k=1}^{N} f_k\,e^{-i2\pi n k/N}. \tag{33}$$

For simplicity, we will define the discrete Fourier transform from time to angular frequency as eq. (33). When computing the Fourier transform from eq. (33), the quickest method is

known as a fast Fourier transform (FFT) and requires that the number of steps N be a power of 2. In our calculations, we use a Cooley-Tukey FFT algorithm [60] with N = 1024. The requirements for calculating eq. (30) within this FFT algorithm are, therefore, a complex array of the calculated values of the time dependent autocorrelation function truncated to the second term. Rewriting eq. (29) using Euler's relation yields [38]

$$\langle e^{i\omega(\mathbf{R})t} \rangle = \text{Re}\,\langle e^{i\omega(\mathbf{R})t} \rangle + \text{Im}\,\langle e^{i\omega(\mathbf{R})t} \rangle, \tag{34}$$

where the real and the imaginary parts are given by

$$\text{Re}\,\langle e^{i\omega(\mathbf{R})t} \rangle = \exp\left[\text{Re}\,(A_1(t) + A_2(t))\right]\cos\left[\text{Im}\,(A_1(t) + A_2(t))\right],$$

$$\text{Im}\,\langle e^{i\omega(\mathbf{R})t} \rangle = \exp\left[\text{Re}\,(A_1(t) + A_2(t))\right]\sin\left[\text{Im}\,(A_1(t) + A_2(t))\right]. \tag{35}$$

In eq. (35),

$$\text{Re}[A_1(t) + A_2(t)]$$

$$= 4\pi\rho_P \int_0^{\infty} dr\, r^2\, g_{PD}(r)\,[\cos(\Delta V(r)\,t) - 1]$$

$$+ 4\pi\rho_P^2 \int_0^{\infty}\int_0^{\infty} dr_1\, dr_2\, h(r_1, r_2)\,[\cos(\Delta V(r_1)\,t)\,\cos(\Delta V(r_2)\,t)$$

$$+ 1 - \sin(\Delta V(r_1)\,t)\,\sin(\Delta V(r_2)\,t)$$

$$- \cos(\Delta V(r_1)\,t) - \cos(\Delta V(r_2)\,t)], \tag{36}$$

And

$$\text{Im}[A_1(t) + A_2(t)]$$

$$= -4\pi\rho_P \int_0^\infty dr\, r^2\, g_{PD}(r)\, [\sin(\Delta V(r)\,t)]$$

$$- 4\pi\rho_P^2 \int_0^\infty \int_0^\infty dr_1\, dr_2\, h(r_1, r_2)\, [\sin(\Delta V(r_1)\,t)\,\cos(\Delta V(r_2)\,t)$$

$$+ \cos(\Delta V(r_1)\,t)\,\sin(\Delta V(r_2)\,t)$$

$$- \sin(\Delta V(r_1)\,t) - \sin(\Delta V(r_2)\,t)]\,, \tag{37}$$

With

$$h(r_1, r_2) = r_1\, g_{PD}(r_1)\, r_2\, g_{PD}(r_2) \int_{|r_1 - r_2|}^{|r_1 + r_2|} s\, [g_{PP}(s) - 1]\, ds\,.$$

The output of the FFT is a complex function of frequency. The real portion of this complex function is obtained and then convoluted with a standard Gaussian slit function. The final output is the simulated line shape function. Since eqs. (29), (36) and (37) depend on both the radial distribution functions and the ground-state and excited-state intermolecular potentials, these are discussed in more detail below.

RADIAL DISTRIBUTION FUNCTION

After significant investigation, we found that the most stable calculation technique for obtaining radial distribution functions for this problem was the analytical solution of the Ornstein-Zernike relation within the Percus-Yevick (PY) closure [59]. Although this solution for a binary system yields four coupled integro-differential equations, dilute solutions (i.e., $\rho_D \ll \rho_P$) allows these equations to be reduced to the calculation of the perturber/dopant radial distribution function $g_{PD}(r)$ and the perturber/perturber radial distribution function $g_{PP}(r)$. This solution is given by [12, 59]

$$g_{PD}(r) = r^{-1}\, e^{-\beta V_g(r)}\, Y_{PD}(r)$$

$$g_{PP}(r) = r^{-1}\, e^{-\beta V_g'(r)}\, Y_{PP}(r)\,, \tag{38}$$

Where

$$Y_{PD}(r) = \int_0^r dt\, \frac{dY_{PD}(t)}{dt}\,,$$

$$Y_{PP}(r) = \int_0^r dt\, \frac{dY_{PP}(t)}{dt}\,, \tag{39}$$

With

$$\frac{d}{dr}Y_{PD}(r) = 1 + 2\pi\rho_P \int_0^\infty dt \left(e^{-\beta V_g(t)} - 1\right) Y_{PD}(t)$$

$$\times \left[e^{-\beta V_g'(r+t)} Y_{PP}(r+t)\right.$$

$$\left. - \frac{r-t}{|r-t|} e^{-\beta V_g'(|r-t|)} Y_{PP}(|r-t|) - 2t\right],$$

(40)

$$\frac{d}{dr}Y_{PP}(r) = 1 + 2\pi\rho_P \int_0^\infty dt \left(e^{-\beta V_g'(t)} - 1\right) Y_{PP}(t)$$

$$\times \left[e^{-\beta V_g'(r+t)} Y_{PP}(r+t)\right.$$

$$\left. - \frac{r-t}{|r-t|} e^{-\beta V_g'(|r-t|)} Y_{PP}(|r-t|) - 2t\right],$$

and with V_g and V'_g being the ground state perturber/dopant and ground state perturber/perturber intermolecular potentials, respectively.

Intermolecular potentials

Eqs. (29), (34) - (40) are explicitly dependent on the excited-state and ground-state perturber/dopant intermolecular potentials through $\Delta V(r)$, and are implicitly dependent

on the perturber/perturber and perturber/dopant ground-state intermolecular potential via $g_{PP}(r)$ and $g_{PD}(r)$. Thus, these simulations require one to develop a single set of ground-state and excited-state intermolecular potential parameters for each system. A standard Lennard-Jones 6-12 potential, or

$$V(r) = 4\varepsilon\left[\left(\frac{\sigma}{r}\right)^{12} - \left(\frac{\sigma}{r}\right)^6\right],$$

(41)

was chosen for the atomic perturber/perturber ground-state intermolecular interactions, and non-polar dopant/perturber ground-state intermolecular interactions. The ground-state molecular perturber/perturber intermolecular potential was a two-Yukawa potential, or

$$V(r) = -\frac{\kappa_0 \varepsilon}{r}\left[e^{-z_1(r-\sigma)} - e^{-z_2(r-\sigma)}\right].$$

(42)

The ground-state polar dopant/perturber intermolecular interactions were modeled using a modified Stockmeyer potential

$$V(r) = 4\varepsilon' \left[\left(\frac{\sigma'}{r} \right)^{12} - \left(\frac{\sigma'}{r} \right)^{6} \right] - \frac{1}{r^6} \alpha_P \mu_D^2 ,$$

(43)

which can be rewritten in standard Lennard-Jones 6-12 potential form [12], with

$$\varepsilon = \varepsilon' \left[1 + \frac{\alpha_P \mu_D^2}{4\varepsilon' \sigma'^6} \right]^2 ,$$

$$\sigma = \sigma' \left[1 + \frac{\alpha_P \mu_D^2}{4\varepsilon' \sigma'^6} \right]^{-1/6} .$$

(The modified Stockmeyer potential includes orientational effects via an angle average that presumes the free rotation of the polar dopant molecule.) An exponential-6 potential, given by

$$V(r) = \frac{\varepsilon}{1 - (6/\gamma)} \left\{ \frac{6}{\gamma} e^{\gamma(1-x)} - x^{-6} \right\} ,$$

(44)

was chosen for the excited-state dopant/ground-state perturber interactions. In eqs. (41) - (44), ε is the well depth, σ is the collision parameter, α_p is the perturber polarizability, μ_D is dopant dipole moment, $\chi \equiv r/r_e$ (where re is the equilibrium distance), and γ is the potential steepness

The Lennard-Jones parameters for the atomic fluids and the modified Stockmeyer parameters for the CH_3I/A_r and CH_3I/K_r interactions were identical to the parameters used to model accurately the perturber-induced shift of the dopant ionization energy for methyl iodide in argon [12–14, 61], krypton [12, 15, 61] and xenon [16, 61]. The parameters κ_0, z_1, z_2, ε and σ in eq. (42) were adjusted to give the best fit to the phase diagram of the perturber [61].The parameters ε, σ, χ and γ in eq. (44) were adjusted by hand to give the "best" fit to the experimental absorption spectra of the dopant low-n Rydberg states in each of the fluids investigated here

Xe $6s$ / Ar

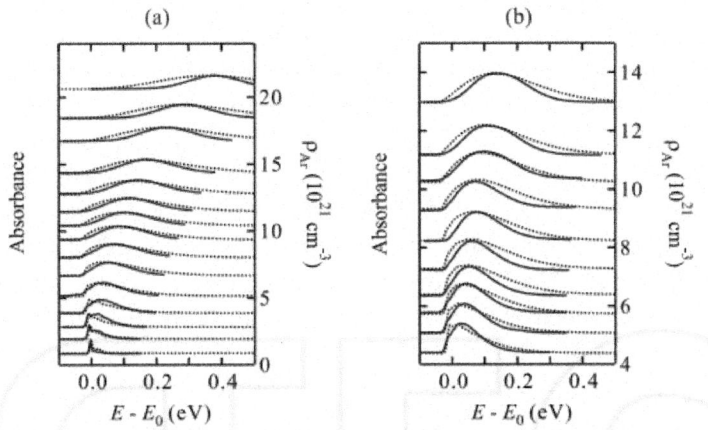

Figure 8: Selected photoabsorption spectra (–, relative scale) and simulated line shapes (···) for the X_e 6s Rydberg transitions at (a) non-critical temperatures and (b) on an isotherm (i.e., $-121.8°C$) near the critical isotherm. The data are offset vertically by the argon number density ρA_r. The transition energy is $E_0 = 8.424$ eV for the unperturbed X_e 6s Rydberg transition

ATOMIC PERTURBERS

Xe low-n Rydberg states in Ar

Xe absorption

The Xe 6s and 6's Rydberg states (where s and s denote the $J = 3/2$ and $J = 1/2$ angular momentum core state, respectively) were experimentally measured in dense argon. As is discussed in Section 2.4.1, when Xe interacts with Ar, ground and excited state dimers form. These dimers are evidenced by blue satellite bands that arise on the higher energy side of the primary Rydberg transition. The Xe 6s Rydberg transition has two such blue satellite bands corresponding to eqs. (24)-(25), whereas the Xe 6's Rydberg transition has a single blue satellite band corresponding to eq. (24) (cf. Fig. 2) [22–24]. The absence of the ground state Xe Ar dimer to Xe Ar eximer transition [i.e., eq. (25)] for the Xe 6's Rydberg state may be caused by an extremely short lifetime preventing our ability to detect the transition or by the XeAr eximer decomposing during the excitation.

The solid lines in Fig. 8 present selected experimental Xe 6s Rydberg transitions doped into supercrtical argon at non-critical temperatures and along an isotherm near the critical isotherm offset by the argon number density.

(Similar data for the Xe 6's Rydberg transitions are not shown for brevity.) It can be clearly seen that the Rydberg transitions broaden as a function of the argon number density. The maximum of the absorption band also shifts first slightly to the red and then strongly to the blue, similar to the original observations of Messing, et al. [25, 26]. Since the ground state interaction between Xe and Ar (or XeAr and Ar) is attractive, the ground states are stabilized by the argon solvent shell. The slight red shift observed at low argon number densities indicates that the xenon excited states (either Xe* or Xe*Ar) are also stabilized by the argon solvent shell. As the density increases, however, argon begins to shield the optical electron from the xenon cationic core, thereby decreasing the binding energy of the optical electron . Thus, as the density of argon increases the energy of the excited state also increases, leading to a blue shift in the transition energy at higher densities. Although not shown, the overall blue shift of the 6s Rydberg transition band is much larger than that of the 6's band at the triple point liquid density of argon. This difference

in overall shift is caused by the difference in the core state of the cation, since the $J = 1/2$ core state has a permanent quadrupole moment. This permanent quadrupole moment increases the interaction of the cationic core and the optical electron, thereby implying that the optical electron is less perturbed by the argon solvent shell. However, since the blue satellite bands also broaden and shift with increasing argon density, the primary Xe transition becomes indistinguishable at medium to large argon number densities. Thus, the argon-induced energy shift of the primary Xe transition cannot be investigated directly using these data. Therefore, to probe perturber critical effects on the dopant excited states, we must first accurately simulate the absorption spectra over the entire argon density range at non-critical temperatures and on an isotherm near the critical isotherm.

Discussion In order to simulate accurately the absorption spectra at high density, any line shape simulation has to include the primary transition, denoted a in Fig. 2 and given by eq. (23), as well as the two XeAr dimer transitions that yield the blue satellite bands, denoted b and c in Fig. 2 and given by eqs. (24) and (25), respectively. For the simulation of Xe in Ar, we chose to use eq. (41) for the ground state Ar/Ar, Xe/Ar, and XeAr/Ar interactions and eq. (44) for the Xe*/Ar and Xe*Ar/Ar interactions. We also required that the simulation use a single set of intermolecular potential parameters for the entire argon density range at non-critical temperatures and along the critical isotherm. All intermolecular potential parameters except the Ar/Ar ground state parameters were adjusted by hand to give the best simulated line shape in comparison to the experimental data. The values of these parameters are given here in Appendix A [37, 40]. The relative intensities of the simulated bands were set by comparison to the absorption spectra of Xe doped into argon at argon number densities where all bands could be clearly identified. Experimentally, at low argon number densities, the ratio of heights between the b band and the primary transition is 0.2 for both the Xe 6s and 6s Rydberg states in Ar. For the Xe 6s Rydberg state in Ar, the ratio of heights between the c band and the primary transition is 0.45. Although for concentrated Xe systems, the ratio of heights for the blue satellite bands to the primary transition would increase with decreasing temperature or increasing perturber number density, this is not the case for the very dilute Xe/Ar system investigated here (i.e., [Xe] < 10 ppm for all argon number densities). Therefore, we can assume that the intensity ratio of the blue satellite bands to the primary transitions stays constant at different temperatures and different argon densities. The dotted lines in Fig. 8 are the simulated line shapes for the Xe 6s transition at non-critical temperatures (cf. Fig. 8a) and on an isotherm ($-121.8°C$) near the critical isotherm (cf. Fig. 8b). A similar figure for the Xe 6's transition is not shown for brevity.

Clearly, the simulated spectra closely match the experimental spectra for all densities. Both the simulated and experimental line shapes show a slight red shift at low argon number densities, followed by a strong blue shift at high argon number densities. With these accurate line shape simulations, moment analyses can be performed on the primary transition in order to investigate perturber critical point effects, as well as to discuss trends in solvation of different dopant electronic transitions in the same simple atomic fluid.

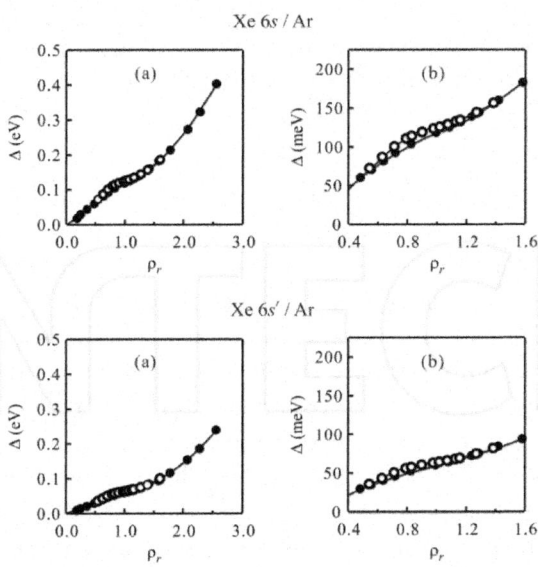

Figure 9: (a) The argon induced shift $\Delta(\rho_{Ar})$, as approximated by eq. (21), of the primary transition for the Xe 6s and 6s Rydberg states as a function of argon number

density ρ_{Ar} at ($\bullet$) non-critical temperatures and ($\circ$) along an isotherm near the critical isotherm. (b) An expanded view of $\Delta(\rho_{Ar})$ The solid lines are a visual aid. See text for discussion.

A line shape analysis was performed on the accurate simulations of the primary Xe 6s and 6s Rydberg transitions in order to determine the average argon induced shift $\Delta(\rho_{Ar})$ of the primary transition, as approximated from the first moment [i.e., eq. (21)]. This moment analysis is shown in Fig 9 as a function of reduced argon number density ρ_r where $\rho_r = \rho_{Ar}/\rho c$ with $\rho_c \equiv 8.076 \times 10^{21}$ cm^{-3} [56]. Fig 9b shows an enhanced view of the perturber critical region with a critical effect in $\Delta(\rho_{Ar})$ clearly apparent. The absence of the red shift observed by Messing, et al. [25, 26] (cf. Fig. 3) results from our performance of a moment analysis on a blue degraded band, instead of a direct non-linear least square analysis using a Gaussian fit function on the primary transition. In other words, while the peak of the primary transition red shifts slightly at low argon densities, the first moment of the band does not, due to the perturber induced broadening

General trends emerged in the behavior of the simulated line shape as a function of the intermolecular potential parameters. For instance, we observed that the strength of the asymmetric blue broadening of a band increases with increasing $\Delta r_e \equiv r_e^{(g)} - r_e^{(e)}$ [where $r_e^{(i)}$ is the equilibrium dopant/perturber distance for either the ground state dopant (i = g) or the excited state dopant (i = e)]. However, the overall perturber-induced energy shift of the band depended on the ground state intermolecular potential well depth $\varepsilon^{(g)}$ as well as Δ_{re}. The slight red shift at low perturber number densities, however, was controlled by the excited state intermolecular potential well depth $\varepsilon(e)$. Comparison of the Xe 6s and 6's transition in Ar

shows that the 6s Rydberg state broadens and shifts to higher energies more quickly than does the 6s state. Since both transitions are excited from the same ground state (implying that the ground state intermolecular potential parameters remain unchanged), Δ_{re} must decrease and $\varepsilon(_e)$ increase in order to simulate the Xe 6s Rydberg state in argon correctly. These general trends proved helpful when determining the intermolecular potential parameters for new systems. Messing, et al. [25, 26] concluded that the argon induced energy shift is density dependent and temperature independent. However, both our experimental absorption spectra and the line shape simulations show a distinct temperature dependence near the argon critical point. To test the sensitivity of the perturber critical point effect, we extracted the perturber dependent shift $\Delta(\rho_{Ar})$ of the simulated primary X$_e$ 6s transition in supercritical argon near the critical density along three different isotherms (i.e., T_r = 1.01, 1.06 and 1.11, where $T_r = T/T_c$ with Tc = −122.3°C). These data are shown in Fig. 10a and clearly indicate that the critical effect is extremely sensitive to temperature and can be easily missed if the temperature of the system is not maintained close to the critical isotherm.

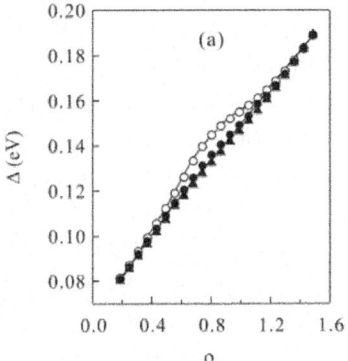 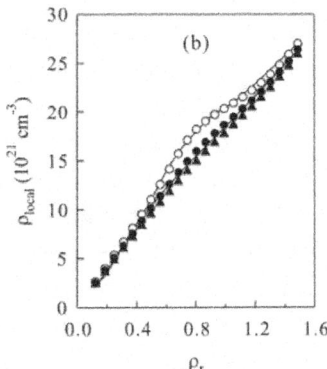

Figure 10: (a) The calculated argon induced shift $\Delta(\rho_{Ar})$ doped into supercritical argon plotted as a function of reduced argon number density at a reduced temperature $T_r =$ 1.01 (⚬), 1.06 (•) and 1.11 (). (b) The local densities ($\rho_{local}1 = g_{max} \rho_{bulk}$) of the first argon solvent shell around a central Xe atom plotted as a function of reduced argon number density at a reduced temperature T_r 1.01 (⚬), 1.06 (•) and 1.11 ((▲)). The solid lines are provided as visual aid. I

If we return to the line shape equation [i.e., eq. (29)], we observe that the two-body interaction term $A_1(t)$ and the three-body interaction term $A_2(t)$ depend on the difference between the excited state and ground state intermolecular potentials and on the perturber/dopant radial distribution function. Since the potential difference will not depend dramatically on temperature, the critical point effect must be dominated by changes in the perturber/dopant radial distribution function $g_{PD}(r)$. In Fig. 10b, we plot the local density of the first solvent shell as a function of the bulk reduced argon number density on the same three isotherms. The $T_r = 1.01$ isotherm shows a much larger density deviation near the critical density in comparison to the other two isotherms. Thus, the argon induced blue shift is caused by the first perturber shell shielding the cationic core from the optical election. This increase in shielding decreases the binding energy of the electron, thereby increasing the excitation energy

CH$_3$I LOW-N RYDBERG STATES IN A$_R$, KR AND X$_E$

CH$_3$I absorption The CH$_3$I 6s and 6's Rydberg states doped into supercritical argon, krypton and xenon were investigated both experimentally and theoretically [38, 40] from low perturber number density to the density of the triple point liquid, at both non-critical temperatures and on an isotherm near (i.e., +0.5°C) the critical isotherm of the perturber. The CH_3I 6s and 6's Rydberg states show perturber-induced energy shifts and broadening similar to that observed for the Xe low-n Rydberg states in supercritical argon. The peak positions of the absorption spectra shift to the red slightly and then strongly to the blue as a function of perturber number densities. This is similar to the behavior for CH_3I in dense rare gases observed by Messing, et al. [27, 28].

Unlike X_e, which forms heterogenous dimers in argon, the CH_3I/perturber interactions are weaker. Thus, CH_3I does not possess blue satellite bands caused by dimer or excimer formation. However, CH_3I does possess a strong vibrational transition on the blue side of the adiabatic transition. Fig. 4 shows the absorption of both the 6s and 6s Rydberg states of CH_3I and clearly illustrates the vibrational state, which represents the CH_3 group deformation vibrational band v_2. The solid lines in Figs. 11 - 13 represent selected photoabsorption spectra for the CH_3I 6s Rydberg transition doped into supercrtical argon, krypton and xenon, while similar plots for the CH_3I 6s transition are not shown for brevity. Experimental spectra of CH_3I in Xe at number densities between 5.0×10^{21} cm^{-3} and 7.0×1021 cm^{-3} could not be obtained, because of the large density deviation induced by small temperature fluctuations ($\approx 2.0 \times 10^{21}$ cm^{-3} for a 0.001°C temperature change) in this density region.

The experimental absorption of CH_3I low-n Rydberg transitions shows that as the perturber number density increases, the $v2$ vibrational band broadens and shifts until it merges with the adiabatic transition. Therefore, determining the perturber induced shift $\Delta(\rho_p)$ of the adiabatic transition from a simple moment analysis of the spectra presented in Figs. 11 - 13 is not possible, and we must perform an accurate line shape analysis of these data in order to extract $\Delta(\rho_p)$ and investigate the perturber critical effect. However, some qualitative information can be gleaned from Figs. 11 - 13. First, the rate of the broadening and the rate of shift for both the adiabatic transition band and the v_2 vibrational transition band differ dramatically for different perturbers. However, although not shown, the CH_3I 6s and 6's transitions have almost the same perturber induced shift, which differs from the behavior observed for the X_e in A_r system previously presented.

DISCUSSION

Although CH_3I in the rare gases does not form dimers or excimers, the accurate simulation of the low-n Rydberg transitions must include both the adiabatic transition, given by eq. (26) and denoted a in Fig. 4, as well as one quantum of the CH_3 deformation vibrational transition v_2 in the excited state, given by eq. (27) and denoted b in Fig. 4. For all of the simulations presented here, we again chose eq. (41) for the ground-state perturber/perturber intermolecular interactions. All of the ground-state dopant/perturber interactions, on the other hand, were approximated with eq. (43). The excited-state dopant/ground state perturber interactions were again modeled using eq. (44). All intermolecular potential parameters except the A_r/A_r, K_r/K_r, X_e/X_e, CH_3I/Ar, and CH_3I/Kr ground state potential parameters were adjusted by hand to give the best simulated line shape in comparison to our experimental absorption spectra. ((The Ar/Ar, Kr/Kr, Xe/Xe, CH3I/Ar, and CH3I/Kr ground-state   potential parameters used are in accord with those employed in our earlier studies of the quasi-free electron energy in rare gas perturbers [12].) Appendix A gives the values for all intermolecular potential parameters used in the line shape simulations presented here. The relative intensities of the simulated bands were fixed by comparison to the absorption spectra of CH_3I at perturber number densities where all bands (i.e., the adiabatic and vibrational transitions) could be clearly identified. Experimentally, at low perturber number densities the ratio of the vibrational band intensity to the adiabatic transition intensity is 0.22 for both the CH_3I 6s and 6's Rydberg states in all three perturbers.

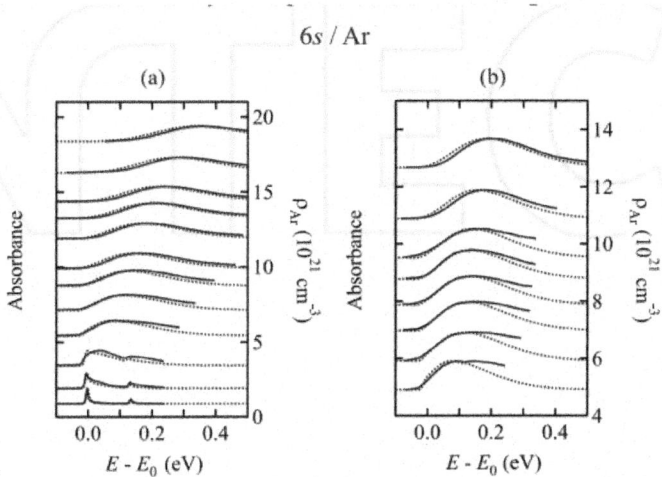

Figure 11: Selected photoabsorption spectra (——, relative units) and simulated line shapes (···) for the CH_3I 6s Rydberg transition in argon at (a) non-critical temperatures and (b) on an isotherm (−121.8°C) near the critical isotherm. The data are offset vertically by the argon number density ρ_{Ar}. The transition energy is $E_0 = 6.154$ eV for the unperturbed CH_3I 6s Rydberg transition. The variation between experiment and simulation is caused by other vibrational transitions and by perturber-dependent lifetime broadening not modeled here.

The dotted lines in Figs.11 - 13 present the simulated line shapes (dotted lines) of the low-n CH_3I Rydberg transitions in the atomic perturbers at non-critical temperatures and on an isotherm near the critical isotherm of the perturber. As was true for X_e in A_r, the simulated spectra closely match the experimental spectra for all densities. Both the simulated and experimental line shapes show a slight red shift at low perturber number densities, followed by a strong blue shift at high perturber densities. Given the accuracy of the simulated line shapes, simulated spectra for CH_3I in X_e in the region where experimental data were unobtainable are also presented in Fig. 13. We should note here that we were able to model the CH_3I 6s and 6's Rydberg states in A_r using the same set of intermolecular potential parameters for both states. This behavior was also observed for the CH_3I 6s and 6s Rydberg states in K_r. With identical potential parameters, the perturber induced shift $\Delta(\rho_p)$ will be the same for the 6s and 6s states. The independence of $\Delta(\rho_p)$ on the dopant cationic core state is different from that observed for X_e low-n Rydberg states in Ar and will be discussed in more detail below. The accurate line shape simulations allow $\Delta(\rho_p)$ for the adiabatic transitions to be extracted using eq. (21).

As with X_e in A_r, the accurate line shape simulations allow a moment analysis to be performed on the CH_3I low-n adiabatic Rydberg transition to obtain the perturber induced shift $\Delta(\rho_p)$ from eq.(21).

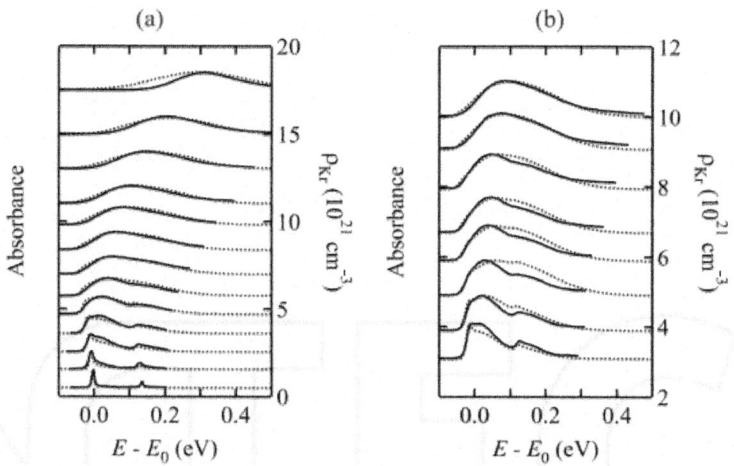

Figure 12: Selected photoabsorption spectra (—, relative units) and simulated line shapes (···) for the CH_3I 6s Rydberg transition in krypton at (a) non-critical temperatures and (b) on an isotherm (−63.3°C) near the critical isotherm. The data are offset vertically by the krypton number density ρKr. The transition energy is $E_0 = 6.154$ eV for the unperturbed CH_3I 6s Rydberg transition. The variation between experiment and simulation is caused by other vibrational transitions and by perturber-dependent lifetime broadening not modeled here.

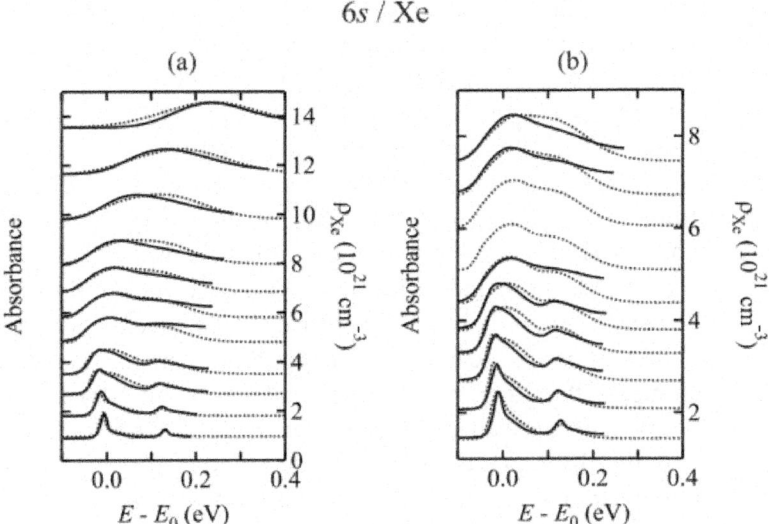

$6s$ / Xe

Figure 13: Selected photoabsorption spectra (—, relative units) and simulated line shapes (···) for the CH_3I 6s Rydberg transition in xenon at (a) non-critical temperatures and (b) on an isotherm (17.0°C) near the critical isotherm. The data are offset vertically by the xenon number density ρXe. The transition energy is $E_0 = 6.154$ eV for the unperturbed CH_3I 6s Rydberg transition. The variation between experiment and simulation is caused by other vibrational transitions and by perturber-dependent lifetime broadening not modeled here.

The first moment of the simulated CH_3I 6s adiabatic transition is plotted as a function of the reduced perturber number density ρ_r in Fig. 14 for the 6s transition. (A similar figure for the 6s transition is not shown for brevity.) The first moment of the simulated adiabatic band does not red shift at low perturber density, as was originally stated by Messing, et al. [27, 28]. This absence of a red shift is again caused by the blue degradation of the adiabatic transition, which places the average energy (i.e., the first moment) of the band to the high energy side of the absorption maximum. The ground state interaction between CH_3I and the perturber is attractive, and therefore the ground state of the dopant is stabilized by

the perturber solvent shell. The slight red shift of the absorption maximum observed at low perturber number densities is indicative of the stabilization of the CH_3I excited states by the perturber solvent shell. As the density increases, however, perturber molecules begin to shield the optical electron from the CH_3I cationic core, thereby increasing the excitation energy of the optical electron. Thus, as the perturber density increases, the energy of the excited state also increases, leading to a blue shift at higher perturber densities

The 6s and 6's Rydberg states correspond to an optical electron in the same Rydberg orbital, but with the cation in a different core state: $J = 3/2$ for s and $J = 1/2$ for s , where J is the total angular momentum of the core. In our investigation of $\Delta(\rho_p)$ for X_e in A_r, we found that $\Delta(\rho_p)$ of the 6s transition is 0.2 eV larger than that for the 6s transition, indicating that the change in the core quadrupole moment affects the dopant/perturber interactions in a dense perturbing medium. However, $\Delta(\rho_p)$ for the CH_3I 6s and 6's Rydberg transitions near the triple point density are identical to within experimental error for the perturbers argon and krypton, and differ only slightly (i.e., 30 meV) for CH_3I in xenon. The insensitivity of these CH_3I/perturber systems to the change in the CH_3I cationic core is probably caused by the large permanent dipole moment of CH_3I, which masks the effect of the quadrupole moment. Xenon, however, is extremely sensitive to electric fields because of its large polarizability. Therefore, the slight difference between the xenon induced shifts of the CH_3I 6s and 6's Rydberg transitions may well be caused by small changes in the permanent dipole moment of CH_3I influencing changes in the induced dipole or local quadrupoles in the xenon perturber.

In the low to medium density range, the energy of the absorption maximum for the 6s and 6's CH_3I Rydberg states has a larger red shift in xenon, which is caused by the larger xenon polarizability. The CH_3I Rydberg states also broaden more quickly in xenon. This increased broadening is probably due to a combination of increased xenon polarizability and an increase in the probability of collisional de-excitation due to the size of xenon. However, $\Delta(\rho_p)$ is larger for argon than for krypton and xenon. This change is caused by an overall decrease in the total number of perturber atoms within the first solvent shell surrounding the CH_3I dopant as the perturber atoms become larger. The variation in the critical point effect, with krypton having a larger effect than argon and xenon, is caused by the strength of the perturber/CH_3I interactions in comparison to the perturber/perturber interactions, coupled with the differences in the ground-state and excited-state dopant/perturber interaction potentials. The CH_3I/K_r ground state potential well depth is close (i.e., 24 K) to the Kr/Kr potential well depth. This implies that the CH_3I/Kr interactions near the krypton critical point will be comparable to the K_r/K_r interactions, thereby leading to a large increase in the local perturber density near the critical point of the perturber, and a larger critical point effect. Similarly, the critical point effect decreases as one goes from krypton to argon to xenon because the difference in well depth for all intermolecular potentials increases.

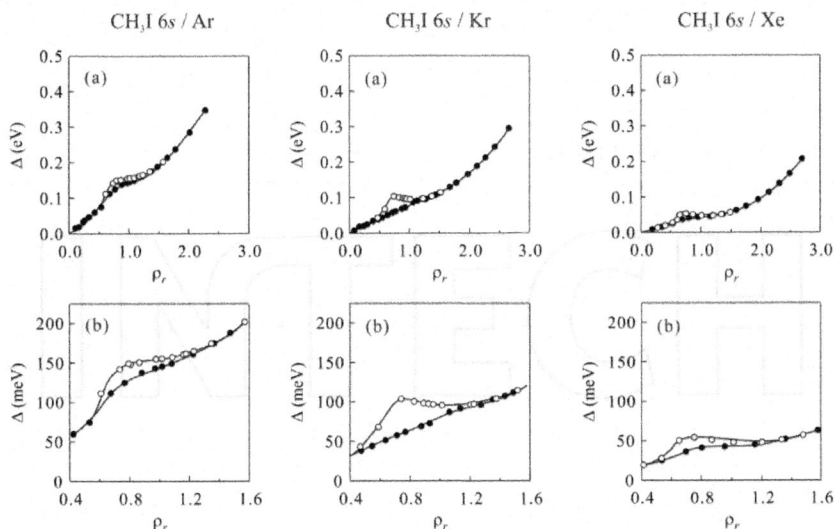

Figure 14: (a) The perturber induced shift $\Delta(\rho_p)$, as approximated by a moment analysis [i.e., eq. (21)], of the simulated primary transition for the CH_3I 6s Rydberg state as a function of the reduced perturber number density ρ_r for argon, krypton and xenon. (•), simulations obtained at noncritical temperatures; (∘), simulations near the critical isotherm. (b) An expanded view of $\Delta(\rho_p)$ near the perturber critical point. $\rho_c = 8.0 \times 10^{21}$ cm^{-3} for argon, $\rho_c = 6.6 \times 10^{21}$ cm^{-3} for krypton and $\rho_c = 5.0 \times 10^{21}$ cm^{-3} for xenon [12]. The solid lines provide a visual aid. See text for discussion.

CONCLUSION

In this work, the structure of low-n Rydberg states doped into supercritical fluids was investigated in several atomic perturbers. Both the experimental absorption spectra and full line shape simulations over the entire perturber density range at non-critical temperatures and along isotherms near perturber critical isotherms were presented for all dopant/perturber systems. These accurate line shape simulations allowed us to extract the perturber-induced energy shift $\Delta(\rho_p)$ from the simulated primary low-n Rydberg transitions. These shifts showed a striking critical point effect in all dopant/perturber systems. Our group also performed similar absorption measurements of atomic and molecular low-n Rydberg states in molecular perturbers [39, 40] with similar results. Because of the brevity of this Chapter, the details of these measurements cannot be presented here. In all of the systems investigated [37–40], the dopant low-n Rydberg states are extremely sensitive to the nature of the perturbing fluid. When these states are doped into supercritical fluids, the surrounding perturbers interact with the central dopant causing shifts both in the dopant ground state energy and in the excited state energy. At low perturber number densities, the dopant/perturber interaction stabilizes the dopant ground state and the low-n Rydberg state. As the perturber density increases, perturber/dopant interactions lead to

the formation of a perturber solvent shell around the dopant core, thereby inducing local perturber density inhomogeneities. This solvent shell begins to shield the optical electron from the cationic core. Therefore, the dense perturber fluid increases the dopant excitation energy, resulting in a blue shift of the abosrption band, which is observed experimentally. The local density of the first perturber solvent shell is almost proportional to the perturber bulk density at non-critical temperatures. However, near the critical isotherm and critical density of the perturber, the dopant/perturber interactions strengthen due to the increased perturber/perturber correlation length. This increased order yields a corresponding increase of the local density in the solvent shell that, in turn, leads to a stronger shielding of the optical electron from the cationic core. Thus, increased blue shifts of the low-n absorption bands are observed in all dopant/perturber systems near the critical point of the perturber. The area of this critical effect is demarcated by the turning points that bound the saddle point in the thermodynamic phase diagram of the critical isotherm. For fluids with similar compressibilities, the structures of low-n dopant Rydberg states in the perturbing fluid show systematic behaviors. At non-critical temperatures, $\Delta(\rho_p)$ is determined by the polarizability and size of the perturbing fluid. The larger the polarizability and, therefore, the larger the size, the smaller the perturber-induced energy shift of the dopant absorption bands. This is caused by the number of atoms that can exist between the optical electron and the dopant cationic core, coupled with the strength of the shielding. The large overall energy shift observed in the dopant low-n Rydberg states perturbed by CF_4 [39, 40], on the other hand, was caused by the larger compressibility of CF_4 in comparison to the other gases in this study [37, 38, 40]. This larger compressibility implies that CF_4 is closer together on average at high perturber number densities than are the other perturbers studied, which increases the local density of CF_4 and, therefore, increases the blue shift in this perturber. The critical point effect, on the other hand, is dominated by the similarity of the perturber/ perturber interaction with the dopant/perturber ground state and dopant/perturber excited state interactions, coupled with the overall local density of the system. In krypton, the well depth of the ground state perturber/ perturber intermolecular potential and the dopant/ perturber intermolecular potential shows greater similarity in comparison to that in A_r and X_e. Moreover, the excited state CH_3I/Kr interaction is slightly stronger than the ground state Kr/Kr interaction. These facts dictate that the largest critical point effect for CH_3I in atomic perturbers is in Kr. Similarly, the largest overall critical effect was observed in CH_3I/CH_4 [39, 40]. This large critical effect is caused by both the ground state and excited state CH_3I/CH_4 interactions having strengths comparable to the CH_4/CH_4 interaction. Although the excited state CH_3I/CF_4 interactions are comparable in strength to the CF_4/CF_4 interactions, the ground state $CH_3I/CF4$ interactions are not close to those of CF_4/CF_4.

Similarly, the X_e/CF_4 ground state interactions are comparable to the ground state $CF4/CF4$ interactions, but the excited state X_e/ground state CF_4 interactions are weaker. Moreover, the bulk critical density in CF_4 is small in comparison to the rest of the perturbers investigated here. This results in the CF_4 critical effect on $\Delta(\rho_p)$ being the smallest one observed [39, 40].

These data sets also allowed us to generate a consistent set of intermolecular potential parameters for various dopant/perturber systems, which are summarized in Appendix A. Several general trends in these parameters can be observed. For atomic perturbers, the steepness of the exponential-6 intermolecular potential (i.e., γ) used to model the dopant excited state/perturber intermolecular interaction decreases with increasing perturber size and polarizability. This trend is reversed in molecular perturbers, were the larger, more compressible CF_4 has a steeper repulsive component in comparison to CH_4. The excited vibrational states of CH_3I always have exponential-6 potentials with a smaller γ in comparison to the CH_3I adiabatic transition in the same perturbing gas. Moreover, the vibrational states always have an equilibrium collision radius that is identical or larger than the collision radius of the adiabatic transition. The excited state collision radii are always larger than the ground state collision radii, as one would expect. However, the interaction strength of the excited state (as gauged by the well depth) can be stronger or weaker than that for the ground state of the same system. These changing interactions are what dominate the variations observed in the critical effects for each of the dopant/perturber systems investigated here.

An understanding of the structure of low-n Rydberg states in supercritical fluids is an important tool in the investigation of solvation effects, since these studies can yield accurate dopant/perturber ground state and excited state intermolecular potentials. We conclude from the present work that the absorption line shapes can be adequately simulated within a simple semi-classical line shape analysis. However, this work focused on highly symmetric perturbers. Future studies should concern more asymmetric perturbers and polar perturbers. Such an extension will require changing the calculation techniques involved in determining the radial distribution functions as well as the type of Fourier transform used to simulate the line shape. Since the excited state is sensitive to the structure of the perturbing fluid, we anticipate that multi-site intermolecular potentials and angular dependent intermolecular potentials will be needed as the perturber complexity increases, in order to model the full line shape accurately.

ACKNOWLEDGEMENTS

All experimental measurements were made at the University of Wisconsin Synchrotron Radiation Center (NSF DMR-0537588), with support from the Petroleum Research Fund (PRF#45728-B6), the Professional Staff Congress - City University of New York, the Louisiana Board of Regents Support Fund (LEQSF(2006-09)-RD-A-33), and the National Science Foundation (NSF CHE-0956719).

REFERENCES

1.   M. B. Robin. Higher Excited States of Polyatomic Molecules Vol. I – Vol. III. Academic Press, New York, 1974, 1975, 1985. and references therein.

2.   R. Reininger, U. Asaf, I. T. Steinberger, and S. Basak. Relationship between the energy v0 of the quasi-free-electron and its mobility in fluid argon, krypton, and xenon. Physical Review B, 28(8):4426–4432, 1983.

3.   R. Reininger, U. Asaf, and I. T. Steinberger. The density dependence of the quasi-free electron state in fluid xenon and krypton. Chemical Physics Letters, 90(4):287–290, 1982.

4.   A. O. Allen and W. F. Schmidt. Determination of the energy level v0 of electrons in liquid argon over a range of densities. Zeitschrift für Naturforschung A, 37(4):316–318, 1982.

5.   W. von Zdrojewski, J. G. Rabe, and W. F. Schmidt. Photoelectric determination of v0-values in solid rare gases. Zeitschrift fur Naturforschung A, 35(7):672–674, 1980.

6.   B. Halpern, J. Lekner, S. A. Rice, and R. Gomer. Drift velocity and energy of electrons in liquid argon. Physical Review, 156(2):351–352, 1967.

7.   W. Tauchert, H. Jungblut, and W. F. Schmidt. Photoelectric determination of v0 values and electron ranges in some cryogenic liquids. Canadian Journal of Chemistry, 55(11):1860–1866, 1977.

8.   J. R. Broomall, W. D. Johnson, and D. G. Onn. Density dependence of the electron surface barrier for fluid helium-3 and helium-4. Physical Review B, 14(7):2819–2825, 1976.

9.   J. Jortner and A. Gaathon. Effects of phase density on ionization processes and electron localization in fluids. Canadian Journal of Chemistry, 55(11):1801–1819, 1977.

10. N. Schwenter, E. E. Koch, and J. Jortner. Electronic Excitations in Condensed Rare Gases. Springer-Verlag, Berlin, 1985.

11. U. Asaf, R. Reininger, and I. T. Steinberger. The energy v0 of the quasi-free electron in gaseous, liquid, and solid methane. Chemical Physics Letters, 100:363–366, 1983.

12. C. M. Evans and G. L. Findley. Energy of the quasifree electron in argon and krypton. Physical Review A, 72:022717, 2005.

13. C. M. Evans and G. L. Findley. Energy of the quasi-free electron in supercritical argon near the critical point. Chemical Physics Letters, 410:242–246, 2005.

14. C. M. Evans and G. L. Findley. Field ionization of c2h5i in supercritical argon near the critical point. Journal of Physics B: Atomic, Molecular and Optical Physics, 38:L269–L275, 2005.

15. Luxi Li, C. M. Evans, and G. L. Findley. Energy of the quasi-free electron in supercritical krypton near the critical point. Journal of Physical Chemistry A, 109:10683–10688, 2005.

16. Xianbo Shi, Luxi Li, C. M. Evans, and G. L. Findley. Energy of the quasi-free electron in xenon. Chemical Physics Letters, 432:62–67, 2006.

17. Xianbo Shi, Luxi Li, C. M. Evans, and G. L. Findley. Energy of the quasi-free electron in argon, krypton and xenon. Nuclear Instruments and Methods in Physics Research A, 582:270–273, 2007.

18. Xianbo Shi, Luxi Li, G. M. Moriarty, C. M. Evans, and G. L. Findley. Energy of the quasi-free electron in low density ar and kr: extension of the local wigner-seitz model. Chemical Physics Letters, 454:12–16, 2008.

19. Xianbo Shi, Luxi Li, G. L. Findley, and C. M. Evans. Energy of the excess electron in methane and ethane near the critical point. Chemical Physics Letters, 481:183–189, 2009.

20. G. D. Mahan. Satellite bands in alkali-atom spectra. Physical Review A, 6:1273–1279, 1972.

21. M. Lax. The franck-condon principle and its application to crystals. Journal of Chemical Physics, 20:1752–1760, 1952.

22. R. Granier, M. C. Castex, J. Granier, and J. Romand. Perturbation of the xenon 1469 a. resonance line by various rare gases and hydrogen. Comptes Rendus de l'Académie des Sciences B, 264:778, 1967.

23. M. C. Castex, R. Granier, and J. Romand. Perturbation of the 1236-a. resonance line of krypton and the 1295-a. resonance line of xenon by various rare gases. Comptes Rendus de l'Académie des Sciences B, 268:552, 1969.

24. M. C. Castex. Absorption spectra of xenon-rare gas mixtures in the far uv region (1150-1500 å): high resolution analysis and first quantitative absorption measurements. Journal of Chemical Physics, 66:3854–3865, 1977.

25. I. Messing, B. Raz, and J. Jortner. Medium perturbations of atomic extravalence excitations. Journal of Chemical Physics, 66:2239–2251, 1977.

26. I. Messing, B. Raz, and J. Jortner. Solvent perturbations of extravalence excitations of atomic xenon by rare gases at high pressures. Journal of Chemical Physics, 66:4577–4586, 1977.

27. I. Messing, B. Raz, and J. Jortner. Perturbations of molecular extravalence excitations by rare-gas fluids. Chemical Physics, 25:55–74, 1977.

28. I. Messing, B. Raz, and J. Jortner. Medium effects on the vibrational structure of some molecular rydberg excitations. Chemical Physics, 23:351–355, 1977.

29. A. M. Halpern. Iterative fourier reconvolution spectroscopy: van der waals broadening of rydberg transitions; the ~b ← ~x (5p, 6s) transition of methyl iodide. Journal of Physical Chemistry, 96:2448–2455, 1992.

30. T. Kalbfleisch, R. Fan, J. Roebber, P. Moore, E. Jacobson, and L.D. Ziegler. A molecular dynamics study of electronic absorption line broadening in high-pressure nonpolar gases. Journal of Chemical Physics, 103:7673–7684, 1995.

31. R. Fan, T. Kalbfleisch, and L. D. Ziegler. A molecular dynamics analysis of resonance emission: optical dephasing and inhomogeneous broadening of ch3i in ch4 and ar. Journal of Chemical Physics, 104:3886–3897, 1996.

32. T. S. Kalbfleisch, L. D. Ziegler, and T. Keyes. An instantaneous normal mode analysis of solvation: methyl iodide in high pressure gases. Journal of Chemical Physics, 105:7034–7046, 1996.

33. S. A. Egorov, M. D. Stephens, and J. L. Skinner. Absorption line shapes and solvation dynamics of ch3i in supercritical ar. Journal of Chemical Physics, 107:10485–10491, 1997.

34. E. Morikawa, A. M. Köhler, R. Reininger, V. Saile, and P. Laporte. Medium effects on valence and low-n rydberg states: No in argon and krypton. Journal of Chemical Physics, 89:2729–2737, 1988.

35. P. Larrégaray, A. Cavina, and M. Chergui. Ultrafast solvent response upon a change of the solute size in non-polar supercritical fluids. Chemical Physics, 308:13–25, 2005.

36. C. N. Tiftickjian and S. A. Egorov. Absorption and emission lineshapes and solvation dynamics of no in supercritical ar. Journal of Chemical Physics, 128:114501, 2008.

37. Luxi Li, Xianbo Shi, C. M. Evans, and G. L. Findley. Xenon low-n rydberg states in supercritical argon near the critical point. Chemical Physics Letters, 461:207–210, 2008.

38. Luxi Li, Xianbo Shi, C. M. Evans, and G. L. Findley. Ch3i low-n rydberg states in supercritical atomic fluids near the critical point. Chemical Physics, 360:7–12, 2009.

39. Luxi Li, Xianbo Shi, G. L. Findley, and C. M. Evans. Dopant low-n rydberg states in cf4 and ch4 near the critical point. Chemical Physics Letters, 482:50–55, 2009.

40. Luxi Li. Atomic and Molecular Low-n Rydberg States in Near Critical Point Fluids. PhD thesis, The Graduate Center of the City University of New York, New York, NY, 2009.

41. S. C. Tucker. Solvent density inhomogeneities in supercritical fluids. Chemical Reviews, 99:391–418, 1999.

42. P. Attard. Spherically inhomogeneous fluids. i. percus-yevick hard spheres: osmotic coefficients and triplet correlations. Journal of Chemical Physics, 91:3072–3082, 1989.

43. P. Attard. Spherically inhomogeneous fluids. ii. hard-sphere solute in a hard-sphere solvent. Journal of Chemical Physics, 91:3083–3089, 1989.

44. J. Zhang, D. P. Roek, J. E. Chateauneuf, and J. F. Brennecke. A steady-state and time-resolved fluorescence study of quenching reactions of anthracene and 1,2-benzanthracene by carbon tetrabromide and bromoethane in supercritical carbon dioxide. Journal of the American Chemical Society, 119:9980–9991, 1997.

45. Y. P. Sun, M. A. Fox, and K. P. Johnston. Spectroscopic studies of p-(n,n-dimethylamino)benzonitrile and ethyl p-(n,n-dimethylamino)benzoate in supercritical trifluoromethane, carbon dioxide, and ethane. Journal of the American Chemical Society, 114:1187–1194, 1992.

46. R. S. Urdahl, D. J. Myers, K. D. Rector, P. H. Davis, B. J. Cherayil, and M. D. Fayer. Vibrational lifetimes and vibrational line positions in polyatomic supercritical fluids near the critical point. Journal of Chemical Physics, 107:3747–3757, 1997.

47. R. S. Urdahl, K. D. Rector, D. J. Myers, P. H. Davis, and M. D. Fayer. Vibrational relaxation of a polyatomic solute in a polyatomic supercritical fluid near the critical point. Journal of Chemical Physics, 105:8973–8976, 1996.

48. R. P. Futrelle. Unified theory of spectral line broadening in gases. Physical Review A, 5:2162–2182, 1972.

49. J. L. Lebowitz and J. K. Percus. Statistical thermodynamics of nonuniform fluids. Journal of Mathematical Physics, 4:116–123, 1963.

50. J. L. Lebowitz and J. K. Percus. Asymptotic behavior of radial distribution function. Journal of Mathematical Physics, 4:248–254, 1963.

51. J. L. Lebowitz and J. K. Percus. Integral equations and inequalitites in theory of fluids. Journal of Mathematical Physics, 4:1495–1506, 1963.

52. M. Born and H. S. Green. A general kinetic theory of liquids. i. the molecular distribution functions. Proceedings of the Royal Society of London Series A, 188:10–18, 1946.

53. R. Kubo and Y. Toyozawa. Application of the method of generating function to radiative and non-radiative transitions of a trapped electron in a crystal. Progress of Theoretical Physics, 13:160–182, 1955.

54. H. C. Jacobson. Moment analysis of atomic spectral lines. Physical Review A, 4:1363–1368, 1971.

55. T. R. Strobridge. The thermodynamic properties of nitrogen from 64 to 300 °k between 0.1 and 200 atmospheres. NBS Technical Note, 129, 1962.

56. A. L. Gosman, R. D. McCarty, and J. G. Hust. Thermodynamics properties of argon from the triple point to 300 k at pressures to 1000 atmostpheres. NBS Technical Note, 27, 1969.

57. W. B. Streett and L. A. K. Staveley. Experimental study of the equation of state of liquid krypton. Journal of Chemical Physics, 55:2495–2506, 1971.

58. W. B. Streett, L. S. Sagan, and L. A. K. Staveley. Experimental study of the equation of state of liquid xenon. Journal of Chemical Thermodynamics, 5:633–650, 1973.

59. E. W. Grundke, D. Henderson, and R. D. Murphy. Evaluation of the percus-yevick theory for mixtures of simple liquids. Canadian Journal of Physics, 51:1216–1226, 1973.

60. W. H. Press, S. A. Teukolsky, W. T. Vetterling, and B. P. Flannery. Numerical Recipes in FORTRAN: The Art of Scientific Computing. Cambridge University Press, New York, 1992.

61. Xianbo Shi. Energy of the Quasi-free Electron in Atomic and Molecular Fluids. PhD thesis, The Graduate Center of the City University of New York, New York, NY, 2010.

Chapter 5

RELATIVIZED QUANTUM PHYSICS GENERATING N-VALUED COULOMB FORCE AND ATOMIC HYDROGEN ENERGY SPECTRUM

Walter J. Christensen Jr.[1,2]

[1]Department of Physics and Astronomy, Cal Poly Pomona University, Pomona, USA

[2]Department of Physics, Cal State Fullerton, Fullerton, USA

ABSTRACT

Though not well-known, Einstein endeavored much of his life to general-relativize quantum mechanics, (rather than quantizing gravity). Albeit he did not succeed, his legacy lives on. In this paper, we begin with the general relativistic field equations describing flat spacetime, but stimulated by vacuum energy fluctuations. In our precursor paper, after straightforward general relativistic calculations, the resulting covariant and contravariant energy-momentum tensors were identified as n-valued operators describing graviton excitation. From these two operators, we were able to generate all three boson masses (including the Higgs mass) in precise agreement as reported in the 2010 CODATA (NIST); moreover local, as-well-as large-scale, accelerated spacetimes were shown to naturally occur from this general relativized quantum physics approach (RQP). In this paper, applying the same approach, we produce an n-valued Coulombs Force Law leading to the energy spectrum for atomic hydrogen, without assuming quantized atomic radii, velocity and momentum, as Bohr did.

INTRODUCTION

In our precursor paper [1], we carried out Einstein's general relativized quantum mechanics approach by re-imagining the basic geometry of local spacetime [2]

. This was accomplished in an analogous way to the how Max Planck solved the blackbody radiation problem—through light quanta absorption and emission. However, instead of a perfect absorber and emitter of light, we proposed a background of fluctuating vacuum energy intermingled with gravitons. Under the principle of general relativity, vacuum energy fluctuations will induce graviton oscillations. Taken together this was enough information to construct a modified flat spacetime metric from normal coordinates describing graviton oscillations [3] . After a straightforward general relativistic calculation on the modified metric, the covariant and contravariant energy momentum tensors emerged as raising and lowering operators describing n-valued graviton excitement. From these operators, we were able to generate all three boson masses (including the Higgs mass) in precise agreement with 2010 CODATA (NIST) [4] ; moreover, accelerated spacetimes were shown to naturally manifest from this approach.

Continuing on with the general relativized quantum physics approach, in this paper, we produce an n-valued Coulombs Force Law, which leads directly to light quanta generating the atomic energy spectrum of hydrogen. We are able to accomplish this without artificially assuming, as Bohr did, quantized: momentum, radii, or velocity for the orbiting electron. Such n-valued atomic energy states emerge naturally from the general relativistic equations (acting on the modified flat spacetime metric), just as Einstein had hoped they would.

ENERGY MOMENTUM OPERATORS

Our general relativistic strategy is to consider flat spacetime at the microscopic level, where vacuum energy fluctuations induce graviton oscillations. Under such a scenario, no longer can the flat spacetime metric be described by the Minkowski metric:

$$g_{\mu\nu} \neq \eta_{\mu\nu}$$

(1)

Instead, we construct a spacetime metric describing graviton oscillations, from normal coordinates [5] , i.e.,

$$g_{\mu\nu} = e^{\sqrt{n}(i\omega)t}\eta_{\mu\nu}, \quad n = 0,1,2,3,\cdots$$

(2)

(Note, at first we did not introduce $\sqrt{n}$ into the spacetime metric. That occurred later when we observed the contravariant energy momentum tensor was describing complex energies phasing cyclically into real and imaginary energies as a function of n . Furthermore, when n=0, or when $2\omega t = n\pi$, spacetime reduces to Minkowski, and results in spacetime nodes both at the local level and then through superposition at the macroscopic level. Whereas the covariant momentum tensor generates only real energy; moreover, the covariant and contravariant operators—acting on the same point in spacetime, display constructive and destructive spacetime interference, as can be understood by:

$$T^{\mu\nu} = g^{\mu\alpha} g^{\nu\beta} T_{\alpha\beta} = e^{-2i\omega t} \eta^{\mu\alpha} \eta^{\nu\beta} T_{\alpha\beta}$$

(3)

This relationship between the mass-energy momentum operators, helps to explain why ordinary matter dominates over antimatter. This is so, because, whereas the covariant energy momentum tensor is always real, the contravariant energy momentum tensor continually phases from real to complex energy-matter. Hence, the energy tensor-operators act destructively or constructively on the same spacetime point to either cancel out antimatter (except under narrow constraints between the phasing), or to add constructively to produce ordinary matter throughout the cosmos.

The following covariant and contravariant n-valued energy momentum tensor-operators were calculated directly from the general relativistic wave equations acting on the modified flat spacetime metric: $g_{\mu\nu} = e^{\sqrt{n}(i\omega)t} \eta_{\mu\nu}$:

$$T_{\mu\nu} = \frac{nc^4}{16\pi G} \begin{pmatrix} -\frac{3}{2}\omega^2 & 0 & 0 & 0 \\ 0 & \frac{1}{2}\omega^2 & 0 & 0 \\ 0 & 0 & \frac{1}{2}\omega^2 & 0 \\ 0 & 0 & 0 & \frac{1}{2}\omega^2 \end{pmatrix}$$

(4)

$$T^{\mu\nu} = \frac{n^2 c^4}{16\pi G} \begin{pmatrix} -\frac{3}{2}\omega^2 & 0 & 0 & 0 \\ 0 & \frac{1}{2}\omega^2 & 0 & 0 \\ 0 & 0 & \frac{1}{2}\omega^2 & 0 \\ 0 & 0 & 0 & \frac{1}{2}\omega^2 \end{pmatrix}$$

(5)

Because ω is the fundamental angular graviton frequency, the n-valued covariant and contravariant energy operators describe graviton excitement. Furthermore, this remarkable n-valued feature of spacetime, is the reason hierarchal particle mass is generated.

We conclude this section, by showing the energy operators obey the conservation of energy:

$$T_{\mu\nu;\nu} = T_\nu^{\mu\nu}; \ \nu = 0$$

(6)

Historically, when this condition was applied to the electromagnetic equations by James Clerk Maxwell, he realized the four electromagnetic equations were not mathematically consistent. To make them consistent (so they obeyed conservation of energy), Maxwell altered Ampere's Law. Upon so doing, he was able to solve one the greatest mystery throughout all time, that of the composition of light. Today, we understand conservation principles provide a powerful tool in ascertaining the laws and workings of natural phenomena. Just as importantly, such consistency conditions provide a means to legitimize (or negate) proposed physical theories. This was indeed the case for classical electromagnetic theory, and now for the general relativized quantum physics approach, we are applying in this paper. The most generalized consistency condition (conservation principle) for gravity interacting with all particle fields, was put into its complete and final form by J. Fang, and is referred to as the Maxwell-Fang consistency condition (MFCC) [6] -[9] .

BOSON MASS

By assuming the two energy momentum operators described above can be detached from the general relativistic wave equations, (without losing their basic spin-like structure), and then applied to more extreme spacetime conditions where gravitons can be excited into higher energy states, in our precursor paper we were able to generate all three boson masses (including Higgs) in precise agreement with CODATA (NIST) [10] . These calculated mass values are as follows:

Z^0-boson mass value:

$$M_{Z^0} = 1.6255(63) \times 10^{-25} \ \text{kg}$$

(7)

Table 1 of the CODATA (2013) reports a Z^0 boson mass of 91.1876(21) GeV/c², where 1 GeV/c² = 1.782661758(44) × 10^{-27} kg; hence the Z-boson mass in kilograms is: 1.6255(66) × 10^{-25} kg. As can be seen, RQP Z-boson mass is in strong agreement with the experimental Z^0 boson mass value.

W-boson mass value:

$$M_{WB} = 1.438(43) \times 10^{-25} \ \text{kg}$$

(8)

The CODATA reports a W-boson mass of: 1.433(32) × 10^{-25} kg, thus we have a second precise match with experiment.

Higgs mass value:

$$M_H = 2.246 \times 10^{-25} \text{ kg} \qquad (9)$$

CERN and others report a Higgs mass of 126.0 GeV or 2.246 × 10^{-25} kg.

All three boson masses calculated from RQP, are in precise agreement with experiment, and so offers confirmation of our general relativized quantum physics approach. Since the Standard Model for particle physics is unable to obtain these hierarchal boson mass values, this is one of the first indications that Einstein's approach to general relativizing quantum physics (RQP), is more fundamental and successful, than is the quantum mechanical approach to nature. In addition to generating boson mass, in our precursor paper [11] , we were able to produce a set of graviton characteristic capable of generating: Planck's constant; speed of light constant; gravitational constant; fine constant structure; electron mass; electric permeability constant, magnetic permeability constant; uncertainty principle, all in precise agreement with CODATA (NIST). The set of four graviton characteristics are denoted by:

$$\left\{ v_g, \; m_g, \; \lambda_g \text{ and } \hat{\omega}_g \right\} \qquad (10)$$

where the graviton characteristic operator $\hat{\omega}$ is given by:

$$\hat{\omega} \equiv n\pi v_g \leq 2\pi \qquad (11)$$

And the graviton frequency, mass and wavelength have values of:

$$m_g = 7.372496678 \times 10^{-63} \text{ kg}, \quad v_g = 1.000000000 \times 10^{-12} \text{ Hz}, \quad \lambda_g = 2.99792458 \times 10^{20} \text{ m}$$
$$(12)$$

In this paper we continue on with RQP approach, and now show Coulombs Force Law becomes n-valued, leading to quantized atomic energy states for the hydrogen atom.

Electron Mass Generated from RQP

After detaching the covariant and contravariant energy momentum operators (calculated on modified flat spacetime), these gravitational operators were then applied to extreme spacetime, whereupon graviton excitations would naturally occur (namely inside the galactic core). The result being, that gravitons represented by the time component in the energy momentum tensor-operator, would switch sign to become repulsive, and so flux from the core at speeds near light. By assuming so, allowed us to calculate the galactic dark halo number density.

$$N_{\text{DMe}} = 1.0 \times 10^{31} \text{ particles/m}^3$$

(13)

By dividing this number density into the relativized energy momentum tensor-operators, yielded the energy per graviton, which we then converted into n-valued boson mass.

In this paper, we follow the same procedure, except rather than generating boson mass, we generate lepton electron mass. To do so means we need to apply a fundamental electron creation frequency to the energy momentum operators. That is to say, we replace the graviton angular frequency by the creation electron-lepton frequency: $\omega_g \Rightarrow \omega_e$ (Note: within RQP all physical values are ultimately generated from the four graviton characteristics discussed previously; and so ω_e serves only as intermediary step to bring clarity to the methodology of RQP). For the covariant energy tensor we have:

$$\frac{T^{00}}{N_D} = n^2 \frac{\left(\frac{3}{2}\omega_e^2\right)\left(\frac{c^4}{16\pi G}\right)}{1.00 \times 10^{31}} = n^2 \left(\psi_{e2}\right) \text{ J/particle},$$

, (14)

(Moreover, RQP allows for n-valued energy momentum operators; this is useful in explaining atomic transition frequencies [12] -[14] and any current discrepancies, however this is not the topic of this paper, and so will not be addressed in any greater detail.) Converting this energy per particle formula into the well known electron mass:

$$m_e = \left(n^2\right)\frac{E}{c^2} = \frac{\left(n^2\right)\psi_e}{c^2} = 9.1093829140 \times 10^{-31} \text{ kg}$$

(15)

Then solving for the coefficient value: ψ_e, we have:

$$\psi_e = \frac{\left(9.1093829140 \times 10^{-31} \text{ kg}\right)c^2}{n^2}$$

(16)

Changing notation for purposes of visual understanding $\psi_e \rightarrow \psi_e^n$, , we solve for the n-value that can generate electron mass: $n = 1$ ground state: $\psi_e^n \rightarrow \psi_e^1$ ground yields the value:

$$\psi_e^1 = \frac{\left(9.1093829140 \times 10^{-31} \text{ kg}\right)\left(2.99792458 \times 10^8 \text{ m/s}\right)^2}{1^2} = 8.187105069 \times 10^{-14} \text{ J}$$

(17)

From this coefficient, the general relativistic contravariant energy momentum tensor-operator, becomes n-valued mass-energy generator for electrons:

$$m_e = \frac{\left(n^2\right) \psi_e}{c^2} = \left(9.1093829140 \times 10^{-31} \text{ kg}\right) n^2$$

(18)

What is being said here, is that, the mass-energy of an electron may become discretely elevated or reduced, due to various spacetime conditions (rather than as Bohr did, by artificially proposing quantized electron angular momentum). This RQP understanding of discrete n-valued changes in the mass-energy of an electron caused by spacetime conditions, is fundamentally congruous to a relativistic understanding—as initially described by special relativity. From a general relativistic approach, if quanta is to manifest, it must do so from causal mass- energy relationships involving spacetime. This is the key feature of RQP; in particular to bound, n-valued electron mass—leading to atomic light quanta emission and absorption.

We now apply n-valued electron mass-energy to Coulombs force law:

$$F_e = k\frac{Zq^2}{r^2} = m_e a = n^2 \left(9.1093829140 \times 10^{-31} \text{ kg}\right) a$$

(19)

where Z is the atomic number of the atom, that we set to $Z = 1$ for hydrogen. Assuming centripetal acceleration, for orbiting electrons, we have:

$$k\frac{q^2}{r^2} = n^2 \left(9.1093829140 \times 10^{-31} \text{ kg}\right) \frac{v^2}{r}$$

(20)

We now solve the n-valued atomic hydrogen energy levels. By rearranging terms, we have:

$$k\frac{q^2}{rn^2} = m_e v^2$$

(21)

This immediately yields the kinetic energy for an electron orbiting about the hydrogen nucleus:

$$K = \frac{1}{2}m_e v^2 = \frac{1}{2}k\frac{q^2}{n^2 r}$$

(22)

By the Virial Theorem the total energy is half the potential energy, and so the total energy is:

$$E_n = K + U = \frac{1}{2}k\frac{q^2}{n^2 r} - k\frac{q^2}{n^2 r} = -k\frac{q^2}{2n^2 r}$$

(23)

This last energy relationship reveals quantized mass leads to n-valued energy states for the hydrogen atom (and in theory, any atom). From a relativity point-of-view, it makes far more sense that atomic spectral energies be derived from n-valued mass-energy, simply because mass and energy are fundamental aspects of Nature; and not with particular electron radii or velocity. Furthermore, RQP is able to provide a theoretical reason as to why discrete energy emission and absorption arise from n-valued mass. This is immediately understood via the relationship between the covariant and contravariant energy momentum tensor-operators, which reveal that the gravitational fields continually undergo constructive and destructive spacetime interference (while generating fundamental n-valued matter):

$$T^{\mu\nu} = g^{\mu\alpha} g^{\nu\beta} T_{\alpha\beta} = e^{-2i\omega t} \eta^{\mu\alpha} \eta^{\nu\beta} T_{\alpha\beta}$$

(24)

This type of interference is different than electromagnetic wave interference; its derivative is spacetime. What this means in regards to electrons bound to the nucleus of a hydrogen atom, is that these electrons are not simply stuck in an orbiting groove. Rather one must try to imagine all the complexity of a collection of hydrogen gas molecules interacting with each other. And in this collection, for each hydrogen molecule, there also occurs electron-proton interaction. All the while, this unimaginable number comprising the hydrogen soup, consists also of electrons moving about with various velocities and orbiting radii. Yet, because of the constructive and destructive interference nature of spacetime itself, all these interacting complexities are expressed in the form of n-valued hydrogen energy absorption and emission spectra. In short, both electron position and velocity lose their significance, whereas n-valued mass becomes the predominant theme in a general relativized quantum physics theory. This means position, charge and the electron constant k, are simply parameters that may be converted to the well known empirical relationship:

$$k\frac{q^2}{2r} \rightarrow hcR$$

(25)

where h is Planck's constant, c is the speed of light, q the charge of an electron, and R is Rydberg's constant (all these constants derivable from the four graviton characteristics). This implies the general relativized atomic hydrogen energy values are then given by:

$$E_n = -k\frac{q^2}{2n^2 r} = -\frac{hcR}{n^2}$$

(26)

CONCLUSION

Following Einstein's General Relativized Quantum approach (RQP), we were able to generate an n-valued Coulombs Law [15] , which led to light quanta and the atomic hydrogen energy spectrum. In terms of relativity, and RQP theory, it makes far more sense to associate n-valued atomic energy levels with mass-energy, rather than with a particular electron radii or orbiting velocity. Furthermore, since general relativity is completely a covariant theory, this implies RQP theory (from which it is founded), has the advantage of providing new spectral predictions manifesting within extreme spacetime conditions, whereas, flat spacetime radiation theories, simply cannot. For example, there is no way for quantum mechanics to make spectral predictions within the galactic core where supermassive black hole exists. However, under the RQP schemata, this determination is possible. In fact, with the RQP energy momentum energy-operators describing constructive and destructive spacetime interference and the formation of matter and antimatter particles, it may help correct Hawking Radiation predictions [16] , so as to be in better agreement with empirical observation. Finally, we mention that we are currently calculating lepton mass from RQP theory. Taken together, RQP theory supports Einstein's vision to have quantum phenomenon emerge from the general relativistic wave equations.

ACKNOWLEDGEMENTS

My foremost respect and gratitude goes to John Fang; a great friend and physicist whom developed Maxwell's consistency formulation into its final and complete form for gravity interacting with all other particle fields. Fang's consistency formulation was applied to the RQP theory presented in this paper, and thus validated the quantized Coulomb's law resulting in the n-valued spectral emission and absorption for atomic hydrogen. I also wish to thank Stephanie Fang for her kindness, hospitality and goodness. Special thanks also goes to Kai Lam, Steven McCauley, Peter Siegel, Antonio Aurilia, Mary Mogge, Alfonso Agnew, Konrad Stein, Jim Feagin, Hedi Fearn and Harvey Leff, for their years in guiding me through the world of physics and mathematics, also thanks goes to Vann Priest in guiding me toward new approaches in teaching physics. Finally, without Natalie Valle and her little Scarlet Rembrandt (artist), life would not have been filled with so many happy and meaningful moments, providing me with the emotional courage and stamina to endure years of reflective thought leading to the General Relativized Quantum Physics approach. Parissa Djafari and Pamela Hope have provided me with much inspiration and world hope. Of course, I am indebted to my mother, Camilla Christensen, who suffered "bowtie Fenster", and all his pet lizards and frogs, as did my grandfather, father and sons Walter and Cleo. Finally, deep respect and admiration goes to Camille Paglia, whose writings concerning art, science and religion, are invaluable to those seeking "deeper meaning".

REFERENCES

1. Christensen, W.J. (2015) General Relativizing Quantum Mechanics (N-Valued Boson Mass). Gravity Research Foundation.

2. Lehmkuhl, D. Einstein's Approach to Quantum Mechanics. Youtube. http://www.youtube.com/watch?v=zbsbc0MfdlE

3. Christensen, W.J. (2007) GERG, 39, 105-110. http://dx.doi.org/10.1007/s10714-006-0360-8

4. Mohr, P.J., Taylor, B.N. and Newell, D.B. (2012) Reviews of Modern Physics, 84, 1527-1605. http://dx.doi.org/10.1103/RevModPhys.84.1527

5. Christensen, W.J. (2014) Manifestation of Dark Energy, Dark Matter, and Planck's Constant Arising from Four Graviton Characteristics. Journal of Gravitation and Cosmology.

6. Fang, J., Christensen, W.J. and Nakashima, M.M. (1996) Letters in Mathematical Physics, 38, 213-216. http://dx.doi.org/10.1007/BF00398322

7. Fang, J. and Fronsdal, C. (1979) Journal of Mathematical Physics, 20, 2264.http://dx.doi.org/10.1063/1.524007

8. Feynman, R. (1962-1963) Lectures on Gravitation. California Institute of Technology.

9. Huggins, E.R. (1962) Quantum Mechanics of the Interaction of Gravity with Electrons: Theory of Spin-Two Field Coupled to Energy. Dissertation Elisha R. Huggins. California Institute of Technology, Pasadena.

10. Mohr, P.J., Taylor, B.N. and Newell, D.B. (2012) Reviews of Modern Physics, 84, 1527-1605. http://dx.doi.org/10.1103/RevModPhys.84.1527

11. Christensen Jr., W.J. (2014) Manifestation of Dark Energy, Dark Matter, and Planck's Constant Arising from Four Graviton Characteristics. In: Melnikov, V., Gravitation and Cosmology.

12. Jentschura, U.D. (2011) Annals of Physics, 326, 516-533.http://dx.doi.org/10.1016/j.aop.2010.11.011

13. Carroll, J.D., Thomas, A.W., Rafelski, J. and Miller, G.A. (2011) The Radius of the Proton: Size Does Matter. T(r)opical QCD II Workshop. AIP Conference Proceedings, 1354, 25-31.

14. Karshenboim, S.G. (2005) Physics Reports, 422, 1-63.http://dx.doi.org/10.1016/j.physrep.2005.08.008

15. Spavieri, G., Gillies, G.T. and Rodriguez, M. (2004) Metrologia, 41, S159-S170.http://dx.doi.org/10.1088/0026-1394/41/5/S06

16. Mersini-Houghton, L. (2014) Physics Letters B, 738, 61-67.

Chapter 6

THE ATOMIC REGULAR POLYHEDRON ELECTRONIC SHELL

Zilong Kong

Wufangfa Computer Technology Service Department, Guangzhou, China

ABSTRACT

The periodic table of elements is arranged based on a series of regular polyhedron. The stability of inert gas atoms can be explained by the distribution of electrons, as well as their motion and magnetic force structure. A magnetic force regular octahedron is proposed. It is a unique configuration that best satisfies the convergence of electrons moving in the same direction within regular polyhedra. In the case of an electrostatic force crust, the formal electron spin accounts for the crusts intrinsic magnetic moment exceeding the speed of light. If one is to consider that the electron has a magnetic outer layer and an electrostatic inner layer, then the question can be solved and abovementioned inference can provide the basis for magnetic force and momentum for the regular octahedron model. The electron periphery has twenty-petal adsorptive substances; the existence of adsorptive substance causes the magnetic force greater than the electrostatic force. Each electronic shell in the regular polyhedron is in accordance with the electron configuration of periodic table of elements; the kinetic track of each electron is a surface of regular polyhedron. The magnetic properties of iron, cobalt, and nickel can be explained by the regular dodecahedron electronic shell of an atom. The electron orbit converged from reverse direction can explain diamond. The adsorptive substances found in atomic nuclei and electrons are defined as magnetic particles called magnetons. The thermodynamic magneton theory can be better explained when it is analyzed using principles of thermodynamics, superconductivity, viscosity, and even in the creation of glass. The structure of the light is a helical line.

INTRODUCTION

The study of quantum mechanics entails quantifying atom from energy and momentum, wave properties of particle and other aspects, and various electron configurations. Based on the structure of hydrogen atom, a single electron system is initially developed. The present method is still confined to the energy level, as in a unitary approach. Solid and liquid do not have characteristic spectrum, and they lacked an item of information in energy difference. This article regards the inert gases atom as the initiation point for multiple electron system. The magnetic force structure, in which the electrons interact, is analyzed for the formation of electron configuration, using the simple geometric model, which provides the obvious details.

The octet rule [1] describes the most appearance probability of electrons of stable inert gases atoms configuration situated at the corners of the cube, and this may regard as which indicates that the electrons are close to resting. Now this article discusses the state at which electrons are moving: if one is to disregard that each electron motion must be around the atomic nucleus, changing its motion to a circle around the line that connects the nucleus with eight vertices of cube, then a configuration of eight electrons with stable motion is gained. Its symmetry is unique in the magnetic force structure.

If the electron has an electrostatic force crust, the formal electron spin accounts for the crusts intrinsic magnetic moment exceeding the speed of light, which was found by Wolfgang Pauli.

Compton wavelength λ_c of electron should be explained by a substance.

THE UNIQUE PROPERTIES OF THE REGULAR OCTAHEDRON CONFIGURATION

When two moving electrons converge, their electrostatic and magnetic forces interact. The electrostatic force produced between the electrons is repulsion; when the direction of motion of two electrons is the same, it brings about an attractive magnetic force. Thus if the direction of motion is the same when two electrons converge, this becomes a special condition, since the stability of the system must be maintained.

When the motion track of electrons is projected around the eight vertices on the six surfaces of cube (regular hexahedron), and the direction of motion of electron is drawn (**Figure 1**), the condition wherein electrons with the same direction of motion converge is satisfied. In the unfurled map, the same number of cube faces denotes synchronization of the track of motion of the electrons projected on these two faces. Because the electron track from an edge comes into a face and comes out from the other edge, the number of entry edges is equal to that of the exit edges, thus the total number of edges of the face must be an even number. All regular polyhedra have only five kinds, with the number of the faces taken separately as 4, 6, 8, 12, 20. This is a mathematical theorem (platonic solids, see Appendix 1). The number of edges of each face in a regular hexahedron is an even number, and other polyhedron is an odd number, thus such structure is unique.

Only the projection of electron motion is reflected above the regular hexahedron. If the circle track of electron motion is to be regarded as a face, then the distances from the nucleus to each of the faces are equal.

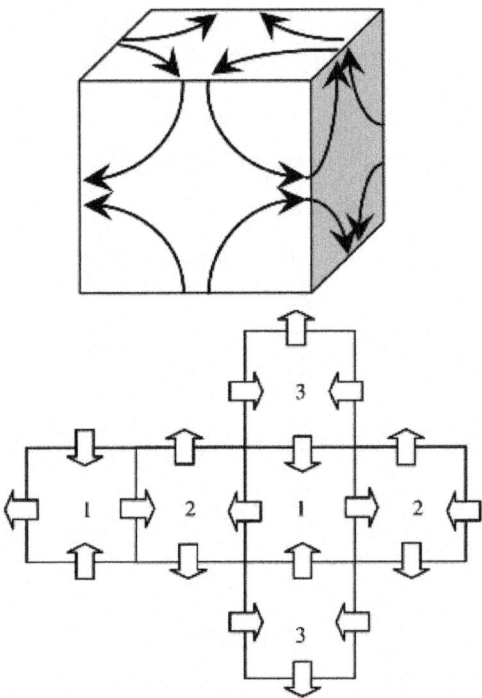

Figure 1: The projection and the unfurled map on the regular hexahedron.

Eight electrons can form a regular octahedron (Figure 2), which only reflects the track of motion of electrons. Electrons converge at the edges of the regular octahedron, satisfying the condition at which electrons converge with the same direction of motion. In the unfurled map, the edges of the same number denote synchronization of the track of motion of electrons at the four edges. Because the electron track from an edge comes into a vertex and comes out from the outer edge, the number of entry edges is equal to that of the exit edges, such that the number of edges of vertex in the regular octahedron must be an even number (see Appendix 1), and other polyhedron is an odd number. This again explains the unique property of the structure. The dual polyhedron of the regular octahedron is a regular hexahedron, which is why this result is obtained.

THE MAGNETIC MOMENT AND ADSORPTIVE SUBSTANCE OF ELECTRON

Usually, the magnetic force between two electrons is far less than the electrostatic force. Their difference is v_e^2/c^2 factor, where v_e is the speed of electron, and c is the speed of light. Then the magnetic force of electrons that converge on a regular octahedron is too small that it can be ignored, but the fact is not of circumstance.

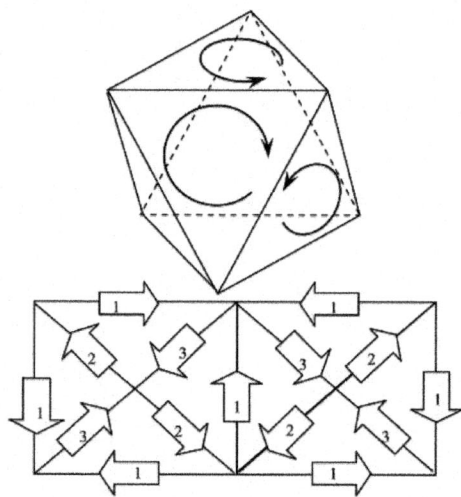

Figure 2: The track on regular octahedron and unfurled map.

The electron has an intrinsic magnetic moment, if the electron is a small sphere with a crust of electrostatic force, where the charge collects on the surface of the sphere and the spinning produces magnetic moment; then the radius of this electrical property sphere is too small, and the intrinsic magnetic moment is too large: if the value of the intrinsic magnetic moment has to be approximated, the surface speed of sphere will have to greatly exceed the speed of light, many literature referred to this problem. Thus, in terms of distance, the magnetic force and the electrostatic force between two electrons can be compared. Otherwise, if some theory treated the electron as a point particle, this issue may be disregarded, or slide over this question.

Now and here assume that the electron has a magnetic outer layer and an electrostatic force inner layer. The electron exterior is surrounded with a layer of magnetic substance that brings about magnetic force; this layer substance is attracted by the electron that joins and rotates about the electron, producing magnetic force and moment. This is the adsorptive substance of electron: its scope radius is a lot greater than the inner layer radius of the electron. When an electron interacts with another particle outside the scope of the adsorptive substance, it produces electrostatic force, magnetic force, and magnetic moment; if another particle enters into the scope of the adsorptive substance, the adsorptive substance is destroyed or changed, the magnetic force and magnetic moment of the adsorptive substance are eliminated or weakened, and the electrostatic force of electron takes over. This assumption is in contrast with that of exceeding the speed of light with a great magnetic radius of the

outer layer of the electron, the magnetic radius r_{em} of electron being the scope radius of the electron adsorptive substance (Figure 3).

The electron magnetic radius r_{em} and the magnetic force of the adsorptive substance can be calculated below. The adsorptive substance charge q moves in a circle with velocity v and radius r_i, the formative magnetic moment μ is equal to current i multiplied by the circumferential area S,

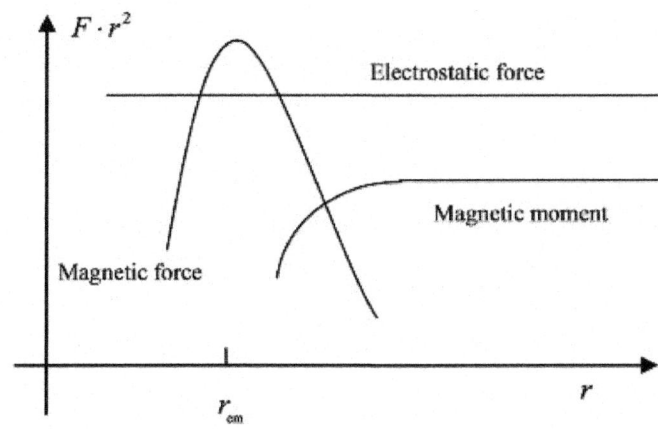

Figure 3: Comparison of electrostatic force, magnetic force and magnetic moment.

$$\mu = iS = q\frac{v}{2\pi r_i}\cdot \pi r_i^2 = \frac{1}{2}qr_iv \tag{1}$$

$$r_i = \frac{2\mu}{qv} \tag{2}$$

The adsorptive substance and the electron have equal quantity charge. If the motion of the adsorptive substance is the same as that of electron spin, it will have an electrostatic force; because the adsorptive substance has only magnetic force but not electrostatic force. Thus the adsorptive substance and electron spin axis are in the same plane, the adsorptive substance symmetrically surrounding the electron side and moving in circle with radius r_a, and tangential to the inner layer of electron (Figure 4). When the inner layer radius of electron is ignored, $r_{cm} \approx 2r_i \approx 2r_a$. If not,

$$r_{cm} = r_i + r_a = \frac{2\mu}{qv} + r_a$$

(3)

Substituted with electron magnetic moment μ_e and electron charge e the speed of the adsorptive substance being the speed of light c, r_a is Compton wavelength λ_c of electron, the magnetic radius of electron can be computed as follows:

$$r_{cm} = \frac{2\mu_e}{ec} + \lambda_C = 7.72766 \times 10^{-13} \, \mathrm{m}$$

(4)

The distance of the adsorptive substance from the electron spin axis is:

$$r_{ca} = r_i - r_a = \frac{2\mu_e}{ec} - \lambda_C = 4.475 \times 10^{-16} \, \mathrm{m}$$

(5)

Using this method one can also obtain the distance of a proton adsorptive substance from a proton spin axis:

$$r_{pa} = \frac{2\mu_p}{ec} - \lambda_C^{proton} = 3.771 \times 10^{-16} \, \mathrm{m}$$

(6)

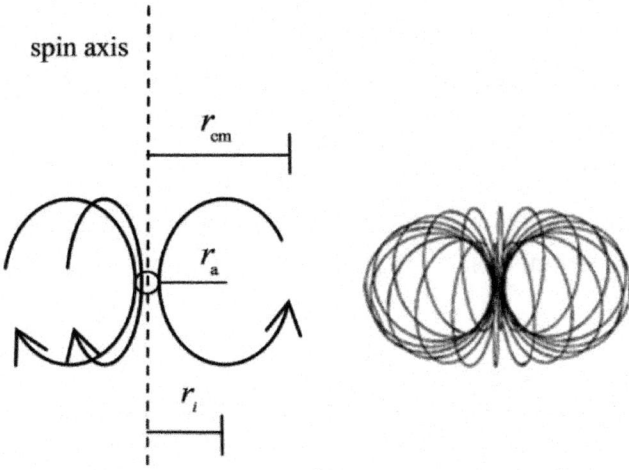

Figure 4: The motion of the electron adsorptive substance.

where r_{ca} is on the same order of magnitude as r_{pa}; from here one can also see that r_a is Compton wavelength λ_c.

Along the circle of the electron spin direction, the ring of the electron adsorptive substance is discontinuous. Suppose P is a positive integer, the adsorptive substance is composed of P circle and the number of petals of electron is P.

The electron magnetic moment reveals the quantity of electron adsorptive substance. The atomic nucleus also has an adsorptive substance that can be released, and the electron outside the nucleus absorbs the adsorptive substance that the nucleus released, eventually gaining an angular momentum. The adsorptive substance that is dispersed as a result of collision is converted into radiated heat.

In a hydrogen atom, the proton charge is equal to the electron; the adsorptive substance generated is equal to the quantity of magnetic moment μ_e, but the proton adsorptive substance is released, generating an electron orbit magnetic moment μ_0 and the proton magnetic moment μ_p. The electron with speed v_e is able to release the adsorptive substance, with the generated magnetic moment $\mu_e v_e / c$. A petal is $\mu_e v_e / pc$; its direction is the same as that of the orbit magnetic moment, tangential to the orbit, and participates in generating orbit magnetic moment μ_0, thus

$$\mu_e = \mu_o + \mu_p - \frac{\mu_e v_e}{pc} \qquad (7)$$

If μ_o is Bohr magneton μ_B, then v_e/c is finestructure constant a, the number of petals of electron P is 20, that is, the electron has twenty petals.

$$\mu_\text{e} - \mu_\text{p} = \mu_\text{B} - \frac{a\mu_\text{e}}{20}.$$
(8)

If the electron absorbs more than one adsorptive substance that the nucleus releases each period, then n is a positive integer

$$n\left(\mu_\text{e} - \mu_\text{p}\right) = \mu_\text{o} - \frac{\mu_\text{e}v_\text{e}}{20c}.$$
(9)

The adsorptive substance release and absorption brings about the transfer of the atomic nucleus spin angle momentum to the electron, becoming the electron angle momentum. The electron and atomic nucleus may keep a determinate distance, thus maintaining the stability of the atom.

THE MAGNETIC FORCE OF ELECTRON ADSORPTIVE SUBSTANCE

After considering the electron adsorptive substance at a position far from the electron, the magnetic force that the electron motion generates is the same v_e^2/c^2 as the electrostatic force. At a position near the magnetic radius of the electron, the speed of the electron adsorptive substance is equal to the speed of light, with the magnetic force being nearly on the same order of magnitude as the electrostatic force of electron. The adsorptive substance does not have electrostatic force, though it has the same quantity of positive and negative charges and possesses polarity. The polarity rotational frequency and phase of the adsorptive substance tangent to electron, is the same. In a hydrogen atom, after coupling between the magnetic moment of the electron adsorptive substance μ_e and that of the electron motion magnetic moment μ_B, and with the proton magnetic moment μ_p with different rotational frequency and phase the relationship is expressed as follows:

$$\sqrt{2}\left(\mu_\text{e} - \mu_\text{B}\right) = \mu_\text{p}.$$
(10)

If all frequencies and phases are the same, the coefficient will be equal to 2. When two electrons with the same velocity and direction converge, their adsorptive substances will also have the same rotational frequencies and phases: the magnetic force of motion of electron will correspond to the charge as that of the electrostatic force. Thus, at the magnetic radius of electron, the total magnetic force corresponding to the electron intrinsic magnetic moment is two times the electrostatic force of the electron. At an appropriate direction, the maximum vector sum of the magnetic force is approximately $\sqrt{2}$ times the electrostatic force. This illustrates the case wherein two magnetic moments or magnetic dipole adjoin. When distance r increases, the magnetic force according to r^{-3} likewise decreases. On a regular octahedron, with the edge serving as a boundary that divides the north and south poles, this on a regular octahedron, with the edge serving as a boundary that divides the north and south poles, this arrangement satisfies the two adjoining poles (**Figure 5**), this requires that the number of edges of each vertical be an even number. Other regular polyhedra cannot satisfy this condition.

The sideways radius of the electron adsorptive substance is r_{em}, and the lengthwise radius is r_a. When the distance of two electrons is greater than $2r_a$, the electrons have the action of both electrostatic force and magnetic moment. When the distance of two electron is near $2r_a$, the magnetic force may be greater than the electrostatic force. When the distance of two electrons is less than $2r_a$, the adsorptive substances are changed, and the electrostatic force is greater than the magnetic force. This property of the electron magnetic force provides a basis for the model of the regular octahedron: if two electrons converge with the same velocity and direction, the magnetic force can be equal to or greater than the electrostatic force.

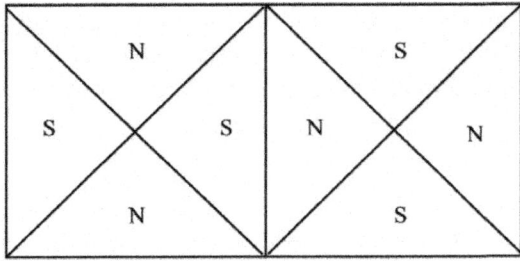

Figure 5: Border upon distribution of north and south poles on regular octahedron.

The magnetic force is two times the electrostatic force, and it also appears in other occasions. The empirical formula of b stability line of atomic nucleus is [2]

$$Z = \frac{A}{1.98 + 0.0155 A^{2/3}}.$$

(11)

Expressing the ratio of the nucleon number A to the proton number Z,

$$\frac{A}{Z} = 1.98 + 0.0155 A^{2/3} \approx \left(2 - \frac{1}{50}\right) + \frac{A^{2/3}}{64}.$$

(12)

The first item explains that at close range, the nuclear force is twice the electrostatic force, the proton number is twice the electron number, 1/50 is caused by the interval of proton and electron. The second item explains that at long distance r, the nuclear force according to r^{-3} decreases, the electrostatic force according to r^{-2} decreases; When the nucleon number A is from 50 to 64, the nuclear force and the electrostatic force at far range are equal, and the atomic nucleus is most stable.

THE REGULAR POLYHEDRON CONFIGURATION

The regular octahedron configuration is the only regular polyhedron in which the electron motions converge in the same direction. If allow the electrons converge in reverse direction, then all regular polyhedra satisfy this requirement. If converge in reverse direction, electrons move to a common edge at different times, there are not collision. It is reasonable to imagine that the electronic shell of the inert gases is at the second shell to outermost shell, and every shell is composed of a regular polyhedron. In each electronic shell, circles of electronic motion are faces of the regular polyhedron; electronic number is the faces number of the regular polyhedron. Two electrons of the first shell compose two parallel rings, with the electron moving in the same direction in a cylinder (**Figure 6**). The spins of two electrons are in the same direction, adjoining north and south poles. They can either be the same or in reverse direction of the orbital angular momentum at one time, thus helium has two configurations. The electrons of each shell of the inert gases element composite configuration and the electron motion modes are as follows.

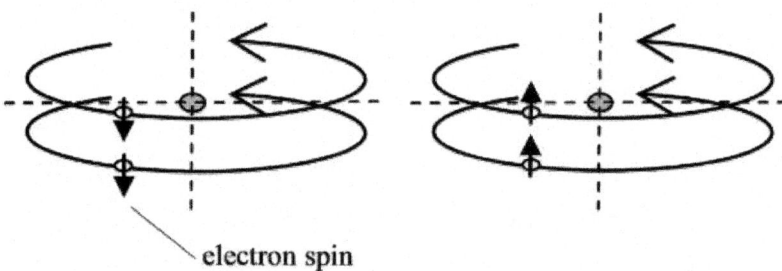

Figure 6: The double ring of same direction and parallel.

First shell: cylinder, same direction and parallel;

Second shell: regular octahedron, converge at same direction;

Middle shells: regular polyhedron, face numbers are 6, 12, 20, allow converge in reverse direction;

Outermost shell: regular octahedron, converge at the same direction.

Thus the electron shell configuration of the inert gases element can be denoted by a row of the face number of regular polyhedron, inasmuch as it is in accordance with the electron configuration of periodic table of elements (Table 1). 18 is dodecahedron and hexahedron, 32 is dodecahedron and icosahedron.

The orbital magnetic moment of the electron also engenders the adsorptive substance. The speed of the same shell electrons is near equation, the electrons can share the adsorptive substance, via its transfer. The energy of each electron tends to equate, and the adsorptive substance even increases the available distance of the magnetic force between electrons. The spin of atomic nucleus extends out and causes the first shell electrons to form a cylinder. For cylinder and regular octahedron, the electron motions of the parallel plane are in the same direction, forming a dipole, consistent with the spin polarity of the atomic nucleus. For hexahedron, dodecahedron and icosahedron, the electron motions of the parallel plane are in the reverse direction, forming a monopole, such as hexahedron with dodecahedron and hexahedron with icosahedron: these combinations can eliminate polarity. The inner shells of electrons are near the atomic nucleus, receive more adsorptive substances that are released by the nucleus, possess a higher negative charge density, and are at the accumulative position of the adsorptive substances. As such, the angular momentum of inner and outer shells is higher, shaping the regular polyhedron with more face numbers. The angular momentum of the middle shells is lower, shaping the regular polyhedron with lesser face numbers.

Before atomic number 21, the regular polyhedron shell and the electron configuration is same in number. The regular polyhedron electronic shells of elements which atomic number are 1 - 103 see Appendix 2.

Table 1: The regular polyhedra that are composed of inert gases electronic shell

Inert element	Electron configuration	Regular polyhedron
He	2	2
Ne	2, 8	2, 8
Ar	2, 8, 8	2, 8, 8
Kr	2, 8, 18, 8	2, 8, 12, 6, 8
Xe	2, 8, 18, 18, 8	2, 8, 12, 6, 12, 6, 8
Rn	2, 8, 18, 32, 18, 8	2, 8, 12, 6, 12, 20, 12, 6, 8

THE CLUSTER OF INERT GAS ELEMENTS

The cluster of inert gas atoms can be described by models represented by regular polyhedrons or cylinders. For the inert gas atoms Ar, Kr, Xe, clusters form according to the preferred magic numbers [3,4].

An explanation using a regular octahedron model can be described as follows: as atoms approach, two closer regular octahedrons have an edge superposed; consequently, two atoms share an edge. Each atom (one regular octahedron) has twelve edges, so each atom can join with another twelve atoms. If twelve atoms are directly set on the edges of a regular octahedron, the resulting structure shows defective tightness. The orbital adsorptive substances on the edge are actually shared by two atoms and these adsorptive substances can undergo excursion. To obtain a tight structure, in a regular octahedron, using a plane that passes through an edge and cutting a vertex with an appropriate angle (Figure 7), twelve edges cut twelve times will result in a regular dodecahedron. If the atoms are then set at its pentagon center, in this way, the resulting structure is the tight, and is in agreement with the descriptions in the literature [4].

A Helium cluster [5] is different from other inert gas elements.

The explanation using the cylinder model is described below: because the helium atom is a cylinder, it can arrange along the direction of a circle and the axis, to form different hexagonal structures, even two layer hexagons (Figure 8), the axes of two layers coincide with each other.

THE MAGNETIC PROPERTIES OF IRON, COBALT, AND NICKEL

Under the frame of quantum mechanics, the origin of metallic ferromagnetism is an issue that is yet to be resolved [6].

The magnetic properties of iron, cobalt, and nickel can be explained through the regular polyhedron electronic shell found in an atom.

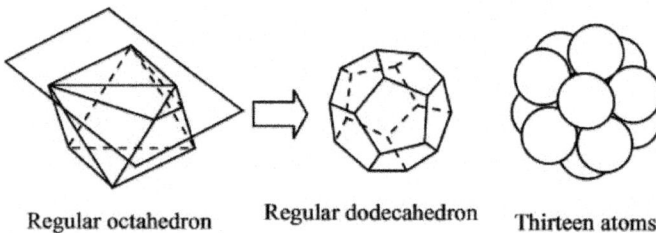

Regular octahedron   Regular dodecahedron   Thirteen atoms

Figure 7: Ar, Kr, Xe cluster.

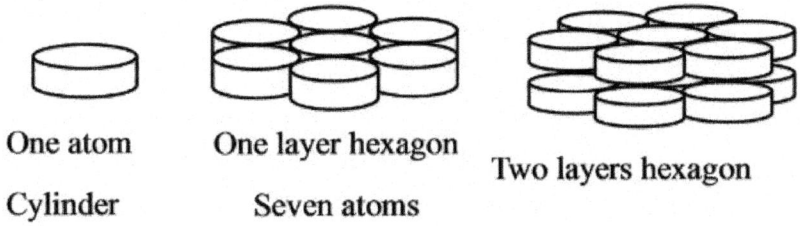

One atom   One layer hexagon

Cylinder   Seven atoms

Two layers hexagon

Figure 8. Helium clusters.

The regular polyhedron electronic-shells of iron, cobalt, and nickel are:

26 Fe   2, 8, 12, 3, 1 2, 8, 12, 2, 2

27 Co   2, 8, 12, 3, 2

28 Ni   2, 8, 12, 5, 12, 8, 12, 4, 2

All of these have a regular dodecahedron; this is the origin of the magnetic properties of iron, cobalt and nickel. The respective free or valence electrons of the outermost shells are 1, 2, 1 or 2, 2, 2; while those of the hypo-outer shell are 3, 3, 5 or 2, 3, 4. The third outer shells are all regular dodecahedrons. Given that hypoouter shell cannot sufficiently form a whole frame. Thus, the hypo-outer shell and third outer shell form a frame of two shells together; the third outer shell of a regular dodecahedron is thereby fixed.

Meanwhile, the motions of electrons on regular dodecahedrons are divided into three cases:

1) On regular dodecahedron, electrons are separated into two hemispheres according to the directions of the electron circumferential motion (Figure 9). At the edges that join the two parts, the directions of the electron motion are the same; there are ten edges which converge with the same direction. Other edges converge from the opposite direction, thereby counteracting the magnetic effect. Ten edges that converge with the same direction engender an equivalent circular electrical current, producing magnetism in the process. Normally, μ is the magnetic moment of the entire regular dodecahedron. In ten pentagons with a total of ten edges, each edge has two electron paths, and μ_e is the magnetic moment of each electron, expressed as:

$$\mu = \frac{1}{5} \times 10 \times 2\mu_e = 4\mu_e$$

(13)

2) On the regular dodecahedron of case 1, there is one pentagon each in the center of the two hemispheres; these are marked as A and B in Figure 10.

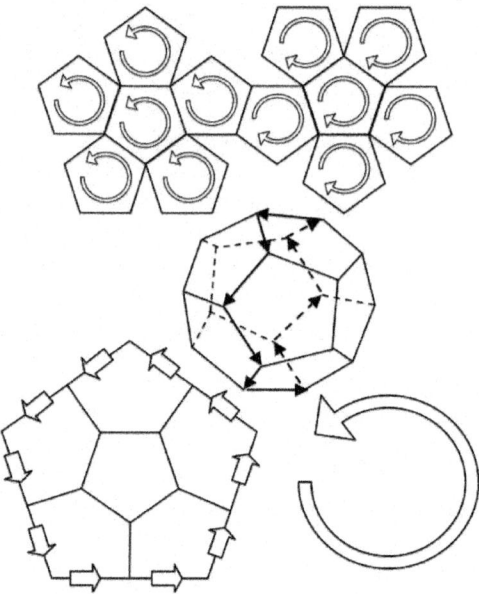

Figure 9: Electron motion and equivalent electrical current in a regular dodecahedron.

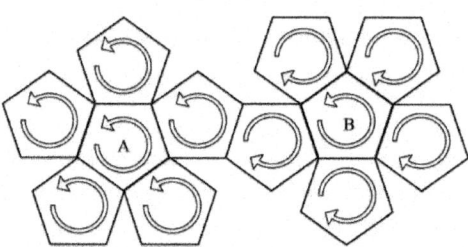

Figure 10: Convergence of the two hemispheres; A, the opposite direction and B, the same direction.

As can be seen, one converges with the same direction, while the other converges from the reverse direction:

$$\mu = 2\mu_e \tag{14}$$

3) Two pentagons in the center of the two hemispheres converge from the same direction as the other edges (**Figure 11**):

$$\mu = 0 \tag{15}$$

If the efficient Bohr magneton numbers of iron, cobalt, and nickel [7] are expressed by $n_B^{Fe} = 2.219$, $n_B^{Co} = 1.715$, $n_B^{Ni} = 0.604$, respectively, then:

$$n_B^{Ni} : n_B^{Co} : n_B^{Fe} \approx 1:3:4 \tag{16}$$

The radii of iron, cobalt, and nickel atoms are very propinquity; thus, the magnetic moment μ_e of each electron are deemed as equal in a regular dodecahedron.

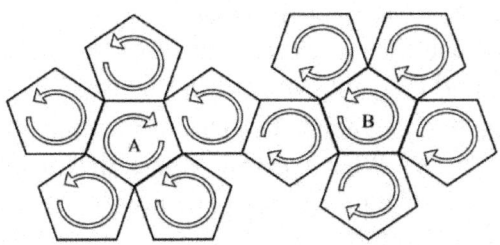

Figure 11: Convergence from the same direction for both A and B.

All atomic structures of iron are case 1:

$$\mu_{Fe} = 4\mu_e \tag{17}$$

A half atomic structure of cobalt is case 1, another half atom is case 2:

$$\mu_{Co} = \frac{4\mu_e + 2\mu_e}{2} = 3\mu_e \tag{18}$$

A half atomic structure of nickel is case 2, another half atom is case 3:

$$\mu_{Ni} = \frac{2\mu_e + 0}{2} = \mu_e \tag{19}$$

From Formulas (17)-(19), we gain:

$$\mu_{Ni} : \mu_{Co} : \mu_{Fe} = 1:3:4 \tag{20}$$

Formulas (16) and (20) are concordant.

Therefore the magnetic properties of iron, cobalt, and nickel can be explained by the regular dodecahedron electronic shell of an atom.

The atom radius of iron is $0.126 \times 10^{-9} m$, from Formula (2) and $2.219\mu_B = 4\mu_e$, in the regular dodecahedron shell, the electronic velocity is $c/289$.

THE THERMODYNAMIC MAGNETON THEORY

The adsorptive substances found in atomic nuclei and electrons are defined as magnetic particles called magnetons.

As a term, "magneton" is commonly used to define or explain a unit of magnetic moment, as in the case of the Bohr and nuclear magnetons, among others. However, in this article, a magneton also denotes a type of particle primarily characterized by a property of magnetism. The magneton is a circle that circumrotates with the speed of light, wherein the radius can change; there are polarities that correspond to the frequency and phase.

The magneton has two different kinds that are separately engendered by either atomic nuclei or electrons; these are called positive and negative magnetons. In an atom, the positive magnetons engendered by the atomic nucleus is released and then arranged into a circulative orbit of the electron, thereby forming an orbit positive magneton (Figure 12).

The negative magnetons engendered by an electron have twenty petals and are attracted by electrons. When electrons move, they tend to release negative magnetons. These negative magnetons (1/20) form orbit negative magnetons, while others (19/20) turn into ray radiation. Some atoms do not have steady orbit negative magnetons.

In this text, the orbit positive magneton is called an orbit magneton for short. The circulative orbit and velocity of electrons determine the orbit magneton quantity, which is capable of maintaining the redundant orbit magneton's ability to transfer to another atom or transform into heat radiation even when orbit and velocity change. This means that the orbit magneton transforms into heat radiation when it does not have a containable orbit. After some orbit magnetons transform into heat radiation, an increase in electronic momentum will occur. Thus, the electronic orbit becomes capable of holding added magnetons, after which it can stop the process and then transform the magnetons into heat radiation. The frequency of radiation of the negative magnetons is generally higher than the positive magnetons (Table 2).

If the perimeter and density of a magneton are $2\pi r_m$ and, respectively, then its mass could be computed by $2\pi r_m D$. The relationship of the polarity periodicity of magneton N_{mt}, the polarity frequency f_m, and radius r_m are expressed as:

$$cf_m = 2\pi r_m / N_{mt}.$$
(21)

The polarity frequencies of a positive magneton in all static atomic nuclei are equal.

The magneton radius, periodicity, and the positive magnetons released from the proton are also equal. However, in the case of positive magnetons, which are released from the atomic nucleus (with more than one nucleon), the magneton radius and periodicity are not necessarily equal.

Meanwhile, just as there are attractions between the negative and positive magnetons, similarly, there are also attractions between the negative and orbit magnetons. The positive magnetons that are attracted by an electron first accumulate at two ends of the electron spin axis; from there, they form an impetus magneton. The impetus magneton may be integrated stepwise to the inner electrons, released out along the motion of the negative magneton of the electron occurring after break up.

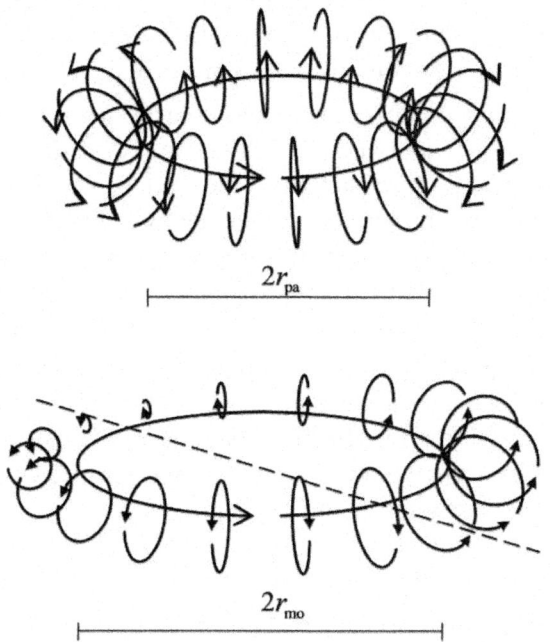

Figure 12: Positive magnetons of proton and orbit positive magneton at electron orbit.

Table 2: Magnetons.

Magneton	Positive magneton	Negative magneton	Orbit magneton	Negative orbit magneton
Polarity	Positive	Negative	Positive	Negative
Location	Proton periphery	Electron periphery	Electron orbit eriphery	Within electron orbit circle
Engendered by (transform from)	Atomic nuclei	Electrons	Positive magneton	Negative magneton
Turn into radiation	Heat low freq.	Ray high freq.	Heat low freq.	Ray high freq.
Magneton radius	$r_{pm} = \lambda_C^{proton} = 2.1031 \times 10^{-16}$ m	$r_{nm} = \lambda_C = 3.8616 \times 10^{-13}$ m	r_{om}	r_{no}
Magneton orbit radius	$r_{pa} = 3.771 \times 10^{-16}$ m	$r_{na} = 4.475 \times 10^{-14}$ m	r_{oo}	r_{no}

It can then cause the electron to obtain impetus. As many positive magnetons accumulate, these compound into a circlet which becomes the compound positive magneton of the large circle. Here, the polarity periodicity of the compound is the average value before the magnetons compound. If the radius of the compounded positive magneton is greater than the diameter of the negative magneton of an electron, then the compound positive magneton shall leave two ends of the electron spin axis and transfer in a parallel position to the side of the electron's negative magneton. This combination of compound positive magneton becomes the orbit magneton. The rules of interaction among positive, compound, and orbit magnetons are explained below (Figure 13):

1) If two orbit magnetons have equal polarity periodicities and radii, the spin axes are in superposition and the spin direction are the same, then there would be no force action; if the spin axes are in superposition but the spin direction is in reverse, then there would be an at traction. On the other hand, if the circles of two magnetons border upon the same plane and the spin direction is the same, then there would be an attraction; in this case, the linear velocity at adjacent point is in reverse direction.

2) If two orbit magnetons have equal polarity periodicities but unequal radii, their spin axes are in superposition, and if the spin direction is the same, then there would be an attraction; but if the spin direction is in reverse, then there would be a repulsion. If the circles of two magnetons border upon the same plane and the spin direction is in reverse, then there would be an attraction; in this case, the linear velocity at adjacent point is the same direction.

3) If two orbit magnetons have unequal polarity peri- odicities, there would be no force action.

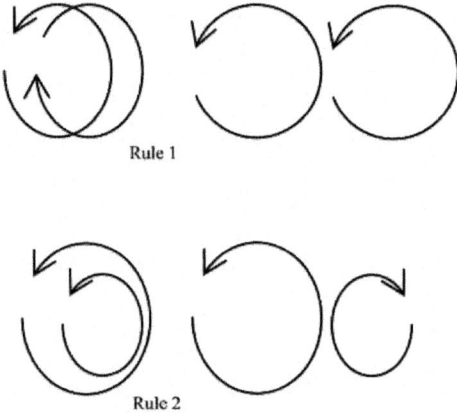

Rule 1

Rule 2

Figure 13: Attraction of orbit magnetons.

If a multi-orbit magneton forms a ring, the circle of each magneton would be perpendicular to the electron orbit (Figure 12). The orbit direction with similar spin directions and arrays would form into a single team or one with a reverse spin direction according to the radii (from large to small). If two magnetons form at the junction and the radii are equal, then the largest or smallest in the team will separate. If the orbit magneton ring on the electron orbit forms 2, 4, 6 and even more teams (two teams at the very least) and the magneton spin directions of the two adjacent teams are in reverse, this will lead to a magnetic moment of zero at the exterior. In Figure 12, the broken line separated two teams.

For the electronic shell of regular polyhedron, the orbit magneton distribute in the edge of regular polyhedron.

If the electronic motion does not periodically turn around, then the orbit magnetons would no longer form a ring so that the beginning would connect to the end. Beside the electron, two teams of orbit magneton would form an olive shape (Figure 14).

When many atoms or molecules come in contact with each other, some parts of the action between electrons are processed via the orbit magnetons. If the orbit magnetons maintain stable attraction, then the object would be in a solid state. The attraction of the orbit magnetons can decrease the distance between electrons as the electron's negative magneton show no apparent attraction. If the distance between electrons is shortened, this would cause a fraction of the orbit magnetons to be pushed out, leading to slippage between atoms or molecules; if this is the case, then the object is said to be in a liquid state. If most of the orbit magnetons are pushed out and the atoms or molecules are also bounced out, then the object is said to be in a gaseous state.

The attraction between the orbit magnetons is in direct ratio with the amount of magnetons. This is also related to the polarity periodicities and radii of such magnetons.

In the gaseous state, the orbit magnetons that are pushed out when molecules come in contact with each other and bounce out become free positive magnetons. The bounced out molecules can also capture free positive magnetons and may become orbit magnetons.

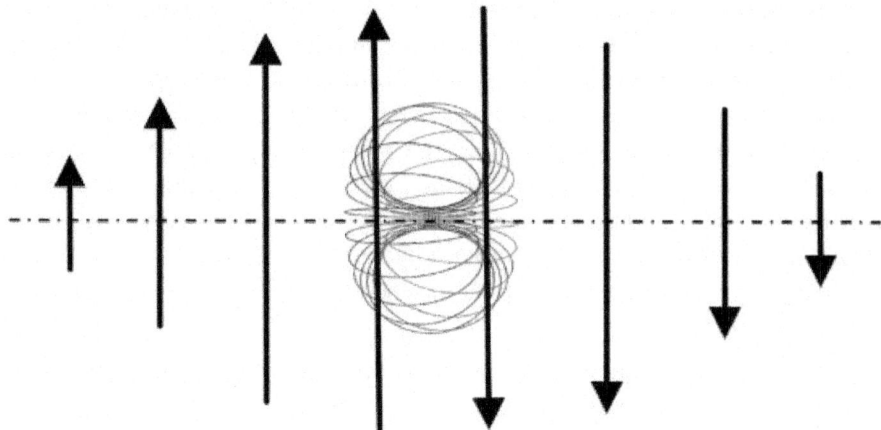

Figure 14. The olive-shaped orbit magneton.

The overall velocity of the free positive magnetons is equal to zero. After the molecule captures the positive magnetons, the speed of the molecule would then be reduced. If molecules have same orbit magnetons, then their speeds are the same.

The action of molecules in the positive magnetons is separated into the force of the free positive magnetons f and the force of the captured orbit magnetons F.

If the attraction of molecules in the free positive magnetons is f, the mass of the free positive magnetons is M, and the acceleration is a, then we have:

$$a = \frac{f}{M}$$

(20)

If the attraction of molecules in the captured orbit magnetons is F; when captured, the attraction is dF, the mass is dM, and the acceleration is the same as the acceleration of the free positive magnetons a, then:

$$dF = adM = \frac{f}{M}dM.$$

(23)

In an equilibrium state, the f of every molecule would be equal. If the molecules with equal dM are considered as the same kind, then $F(M)$ is the distribution of the attraction according to the same kind of molecular total in M stated as:

$$F(M) = \int adM = \int \frac{f}{M}dM = f\ln M.$$

(24)

On the basis of the inter-actional rules of orbit magnetons, only the same kind of molecules may attract each other and push out the orbit magneton; in this situation, the system tends to increase as many different kinds of molecules as possible and capture the maximum positive magnetons.

When molecules bounce out, the work of repulsive force equals the negative work of attraction, expressed by the attraction F multiplied by the maximal radius of orbit magneton r_{max}. Meanwhile, the kinetic energy when molecules bounce out is equal to the work of the repulsive force,

$$E_k = -Fr_{max}.$$
(25)

From Formulas (24) and (25), we derive:

$$E_k = -fr_{max} \ln M, \text{ and} (26)$$

$$M = e^{-E_k/fr_{max}}$$
.(27)

This is the relation of the free positive magnetons mass M and the kinetic energy of molecular E_k. As compared with the Boltzmann factor $e^{-E_k/kT}$, we can see that the Boltzmann constant k is the attraction of molecules for the free positive magnetons f. If it is constant in the equilibrium state, the temperature T is the maximal radius of orbit magneton r_{max}, expressed as:

$$k = f, \text{ and}$$
(28)

$$T = r_{max}.$$
(29)

The dimension of kT is work, the dimension of temperature T is length, and the dimension of k is force. From this relationship, we can say that the magneton theory can work well with thermodynamics.

The maximal radius of the impetus magneton is equal to the minimal radius of the orbit magneton r_{min}, and is also equal to the radius of the negative magneton on static electron r_{em}. The temperature T defined by perfect gas is a macroscopic and statistical quantity of multiunits (molecule); meanwhile, the temperature T of an area near zero K, that is equivalent to: in this area, there are almost no orbit magnetons of the outer shell that may lead to a shift out. Therefore when temperature T nears zero K, the outer shell of most units will have not orbit magnetons; for the few units that have orbit magnetons in the outer shell, the radius of the orbit magneton will take a minimum value as expressed by:

$$T = r_{max} = r_{min} = r_{em} = \lambda_c$$
.(30)

In the outer shell of metal, electrons are far apart from the atomic nucleus, and the number of absorbed positive magnetons could be low such that these would not form into a fixed frame. In such a case, it can rely on the secondary outer shell electrons so that it could be joined together; this is called the hypo-outershell. The outer shell electrons become free electrons in the frame interstice (a free electron has a small quantity of impetus magneton and has an initial kinetic energy E_0). The action between the free electrons is similar to what is occurring in the gaseous state; also, if the free electron has the action of another item from the hypo-outer shell frame via free positive magneton, Formulas (22) and (25) can be changed to:

$$a = \frac{f}{2M}.$$

(31)

$$E_k - E_0 = -2Fr_{max}.$$

(32)

Using a coefficient i denotinge the scale of two item's action, $0 \leq i \leq 2$, and from Formulas (24) and (32), we derive:

$$E_k - E_0 = -ifr_{max} \ln M, \text{ and}$$

(33)

$$M = e^{-(E_k - E_0)/ifr_{max}}.$$

(34)

Suppose m_e is the electron mass, v_e is the electron velocity, v_0 is the initial velocity of electron, the momentum of the action between the electron and hypoouter shell frame is $m_e v_e$, and t is the acting time:

$$m_e v_e = Ft = -ft \ln M, \text{ and}$$

(35)

$$M = e^{-m_e v_e / ft}.$$

(36)

From Formulas (34) and (36), we derive:

$$-(E_k - E_0)/ifr_{max} = -m_e v_e / ft,$$

(37)

$$\frac{\left(\frac{1}{2}m_e v_e^2 - \frac{1}{2}m_e v_0^2\right)}{ifr_{max}} = \frac{m_e v_e}{ft}, \text{ and}$$

(38)

$$t = 2ir_{max}\frac{v_e^2 - v_0^2}{v_e}.$$

(39)

The electrical resistance of metal mostly comes from the action occurring between the electrons and the hypoouter shell frame. It is derived via the transfer of free positive magnetons, electrical resistivity ρ is namely the acting time t, and the maximal radius of the orbit magneton r_{max} is the temperature T, which is treated i as the probability of all collision. Hence, the collision probability between electrons is $P_c = i/2$,

$$\rho = 4P_c T \frac{v_e^2 - v_0^2}{v_e}.$$
(40)

If $v_e < v_0$, or $P_c = 0$, then it is said to be in a superconductive state. The directional motion of the electron can be maintained in the initial kinetic energy E_0. $P_c = 0$ is the non-collision superconductivity, and $v_e < v_0$ the superconductivity when collision speed is less than the initial velocity.

When the temperature T is lower, consider two special cases below:

1)   If P_c and v_0 are close to the constants, then

$$\rho \sim T\left(v_e - \frac{A_1}{v_e}\right).$$
(41)

As such, A_1 is approximately constant. From Formula (33) we derive $T \sim v_e^2$, so

$$\rho \sim T\left(\sqrt{T} - \frac{A_1}{\sqrt{T}}\right).$$
(42)

2)   If v_e and v_0 are close to the constants, the mass of the olive-shaped orbit magnetons would be proportional to the square of the magneton radial; thus, theattraction between electrons would be proportional with the collision probability, $P_c \sim T^2$, leading us to the following:

$$\rho \sim T^3$$
(43)

When the temperature T is higher, then $v_e \gg v_0$. According to the magneton interactional rule, the free electrons attract each other but only if they have the same speed. If there is an increase in free electrons with different speed levels, the range of speed would increase and the collision probability will decrease, $\rho \sim (1/v_e)$, so

$$\rho \sim T.$$
(44)

Therefore, the part of electrical resistivity $\rho > 0$ can disport three segments; these are shown in Formulas (42) to (44). For the expression $\rho \sim T^3$ see also [8].

The high critical temperature superconductivity can be seen in non-collision superconductivity. To increase the critical temperature, the following methods are followed:

1)   Release the free electrons from the atoms of multiple sorts.
2)   The atomic nucleus can release the positive magnetons with different periodicity, these positive magnetons can then form the orbit magnetons of multiple sorts.
3) Increase the speed range of the free electrons.
4) Restrict the motion range of the free electrons.
5) Decrease the density of the free electrons.

In Method 2, the first requirement is an element with a high melting point and the second is an element with low rigidity.

In Formulas (26) to (29), the Boltzmann factor $e^{-E_k/kT}$ is equal to the free positive magnetons mass M. In some liquids, M^{-1} can be viewed as a macroscopic amount referring to viscosity η. In [9,10],

$$\eta = Ae^{Q/RT}.$$
(45)

where Q is the activation energy, R is the molar gas constant, and A is a constant. If one molecule is transformed to many molecules, then k is transformed to R, and E_k is transformed to Q. In these liquids, the viscosity η is inversely proportional to the free positive magnetons mass M in the area:

$$\eta \sim M^{-1}$$
(46)

In [11-13], $\eta = Ae^{B/RT}\left(1+Ce^{D/RT}\right)$, where A, B, C and D are constants, then this is a case when there are two types of free positive magnetons. Fermi distribution is a case when there are two types of attractive particle, Bose distribution is a case when there are two types of repulsive particle.

There is a decrease in free positive magnetons in the process of transformation from liquid to solid state; Its part transform into the orbit magnetons M_o; if their total $M_t = M + M_o$, then E is a constant:

$$\eta = E\left(M_t - M_o\right)^{-1}$$
.(47)

In the presence of orbit magnetons, similar particles can attract each other and molecules and electrons also behave in this manner. In such a case, the ion is also similar. When the glass of oxide (as SiO_2) change from a liquid to solid state, though free positive magnetons decreasing, there are yet enough orbit magneton, the same ions will attract each other and can separately form an ion group of a similar type. The oxygen ions can then attract each other and form a group of oxygen ions; the silicon ions can attract each other and form a group of silicon ions. There are electrostatic attractions between the groups of oxygen and silicon ions. The structure of a glass is formed when a large amount and space of groups of oxygen ions connect and form a net where a group of silicon ions are embedded.

The chemical reaction is regarded as that occurs the electronic transfer, because only the electron those sort and velocity are same attract each other, the course of electron transfer from a sort of atom to another sort of atom, except electrostatic force require also satisfy two condition: first the electronic velocity is same, second is the mix of magneton: the magneton on electron come from several atomic positive magneton, form the orbit positive magneton on electron after mixed. The magneton mix also may be the change of magneton supply, the source of positive magneton from a sort of atom change to another atom. The magneton mix also may the electronic velocity go to equal. Usually the magneton mix easy in high temperature and pressure, and the electronic kinetic energy easy over the electrostatic force, the chemical reaction easy progress. In the situation of temperature and pressure are not change, the magneton mix becomes mostly factor that influence the chemical reaction.

Magneton and Electromagnetic Induction

The magneton of electron formerly was circle. When two electrons approaching to a scope of distance, the magneton interact and being distorted, the circle plane occur bend; if the distance of two electrons are not changed anew, the bend of circle keep and are not change; at the farther position there is away the central electron, on the edge, the plane of magneton rotated a angle, the angle between circles become uneven. Formerly the magneton moving along circle with the velocity of light, this the tangent direction, the magneton is tangent with the central electron also, both spins keep be in mesh in certain phase; the circle of magneton swing in other direction when distortion, the advance speed is not speed of light in tangent direction, but is less than speed of light, the electron must moving backward and keep be in mesh. This backward motion of electron is the electromagnetic induction (Figure 15).

The motion and spin axis direction of electron incline to superposition when the electromagnetic induction. Two electrons appear the electromagnetic induction; firstly the magnetons of electrons attract each other, the spin axis of electrons incline to same direction, and then the magneton swing, the electron moving backward. Two conditions which the free electron can electromagnetic induction is also two sort swings: the swing of the spin axis of electron and the swing of magneton. The swing of the spin axis cause the spin axis of electrons turning to same direction; the swing of magneton cause the electron that keep be in mesh with magneton moving backward. The electromagnetic induction is the change of the magnetic force causing the swing of magneton of electron, the electron moving, this induction differs from the magnetic force (Figure 16).

Two relation of direction are reverse: the relation of the spin axis of electron with the act direction when electromagnetic induction; and the relation of the spin axis with motion direction when thermal motion.

A motion electron comes into magnetic area; firstly the spin axis is turned, the turned place always got behind with interact point, the spin axis incline to same direction with the electron of the back magnet, and then the magneton swing, the electron plus a motion that moving backward along the spin axis, and is the reverse of electron of the back magnet, it is close perpendicular to the motion direction, but is not completely perpendicular to the motion direction.

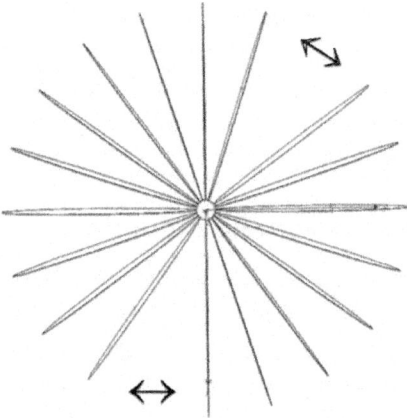

Figure 15. The swing of magneton.

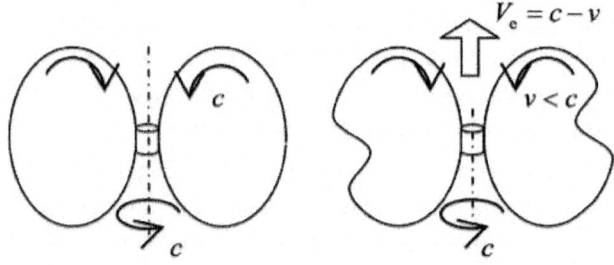

Figure 16. The swing of magnetons causes the electron moving.

A motion electron is affected by magnet, the force that effecting electron f_m with the motion direction of electron (or velocity v_e) forms an obtuse angle ¶ > 90° (Figure 17).

The magnetic induction satisfies three conditions:

1)   The spin axis of electron turning.

2)   A revolving ring with velocity of light exists in electron.

3)   The swing of revolving ring.

The momentums of instantaneous Lorentz force are not conservation, its action force and effect is perpendicular; the electromagnetism induction satisfies three conditions above are momentums conservation.

The delay and inertia of the electromagnetic induction are greater than the energy of ionization, this explain also three conditions above.

The electromagnetic induction can engender voltage, and engender current if there is a loop. When the area of an electric lead circle changed, each part that lead moving engenders voltage, the directions of these voltages are same; the perimeter of circle changed simultaneity, the perimeter has difference, the voltage to the perimeter difference are in direct ratio, the current is that the perimeter difference multiplies the amount of electron, the perimeter difference to the amount of electron are in direct ratio, the area difference to the square of the perimeter difference are in direct ratio, therefore the current to the area difference are in direct ratio.

THE ELECTRONIC SHELL OF CARBON

The electronic shell of the carbolic atom is 2, 4. Four electrons of the outermost shell formed a regular tetrahedron.

In the regular tetrahedron of atom of diamond, all four electrons converge from reverse direction; each atom is connected by superposition of vertex, that is share four vertexes, therefore each atom connect with other four atoms, and is symmetrical (Figure 18).

In the regular tetrahedron of atom of graphite, three electrons is converged from reverse direction, these three electrons is converged from same direction with the other one electron, in six edges of the regular tetrahedron, there is three edges of one face is converged from same direction, other three edges is converged from reverse direction.

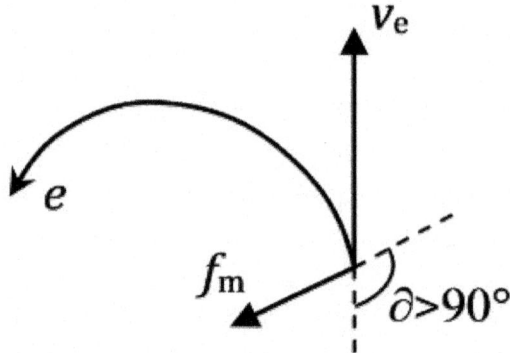

Figure 17. A motion electron is affected by magnet.

Each atom is connected by superposition of three   edges, which are converged from same direction, so each atom is connected with other three atoms, and is in a same plane (Figure 19), these atoms can connect into hexagon.

The converged electron orbit has two sorts of the same or reverse direction from the rule of magneton attract each other: when converge from same direction, the magneton of same radius attract each other; when converge from reverse direction, the magneton of different radius attract each other. The negative magnetons attract each other when converge from same direction, cause the electron orbit close up, and push out the orbit magneton (Figure 20); the negative magnetons do not have attract when converge from reverse direction, electron orbit the do not near mostly, and do not push out the orbit magneton (Figure 21). The electron of atom of diamond all are converge from reverse direction, electrons of adjacent atom are also converge from reverse direction, therefore do not push out the orbit magneton, have some attraction, and equipoise with electrostatic force of nucleus, the electron orbit is most stable.

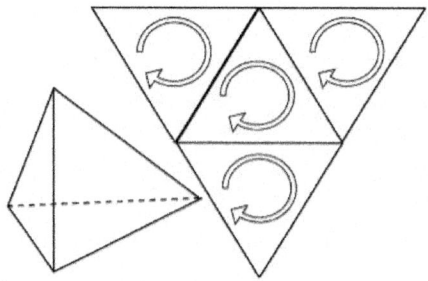

Figure 18. Diamond: four electrons converge from reverse direction.

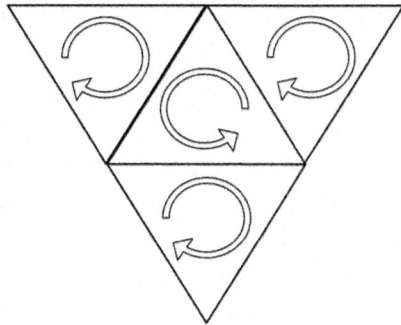

Figure 19. Graphite: three edges of same plane converge from same direction.

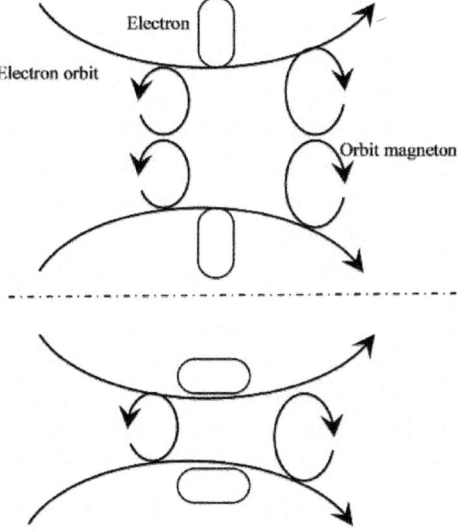

Figure 20. Converge from same direction.

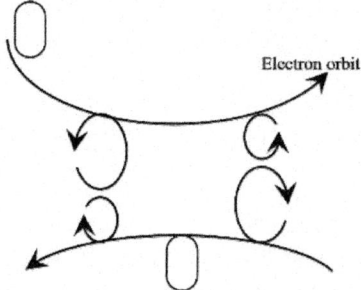

Figure 21. Converge from reverse direction.

THE HELICAL LINE MODEL OF THE LIGHT

The structure of the light is a helical line. The axis of the light helix is same from the direction of motion of light. The direction of motion of light is its direction of travel or propagation direction. β is the helix angle of the light helix, $45° \le \beta \le 90°$. The helix angle approach 45° if the light density is high; when the light density debases, the helix angle increase, maximum approach 90°, the helix almost become a straight line. The wavelength of light is the screw pitch of the helix. If the helix angle of the light helix is already 45° and the light density continue to increase, then the screw pitch of the helix change into short, that is the wavelength of light change into short (Figure 22).

Two planes that are parallel with axis cut symmetrically helix, split helix into three parts, the middle part is a polarized light; if remove the middle part, two remnant parts are also a polarized light.

The light in a magnetic field, the advance have a resistance; the subsequent part of the light helix has an act of pushing for the front part. In the act of resistance and pushing, the front part of the light helix comes into deflection; the deflection of the front part leads the subsequent part and comes into deflection. Therefore the light have a deflection in a magnetic field, the deflection of polarized light can easy be observed.

When a light come close and into an object, have an act first with the magneton of object.

Compton Effect is the result of act of the light and the negative magneton. When light get across the negative magneton, if the axis of the light helix and the spin axis of the negative magneton is superposition, then the helix has a repulsive force, the helix is expanded, the radii of the helix adds the radii of the negativemagneton, that is Compton wavelength λ_c of electron; the screw pitch of the helix adds the perimeter of the negative magneton λ_c, so the wavelength of light adds λ_c.

The orbit positive magneton included two teams of positive magneton with reverse spin, when the light meets the orbit positive magneton, if the axis of the light helix is perpendicular to the circle of the positive magneton, then the light is reflected; if the axis of the light helix is parallel to the circle of the positive magneton, then the light gets across the positive magneton, come into being refraction.

When the velocity of electron change, the radii of the negative magneton of electron decrease, the superfluous substance of the negative magneton become mostly the radiation of light, the few become the orbit negative magneton of electron. Along the straight line if count the frequency of light; along a circle if count the frequency of magneton. A light can have an act with the orbit negative magneton that the frequency is same, and then is absorbed by the atomic nucleus, translate into the positive magneton. The translation about the magneton and the radiation of light or heat express below:

The positive magneton on the atomic nucleus à The orbit (positive) magnetons on the electron orbit à The radiation of heat.

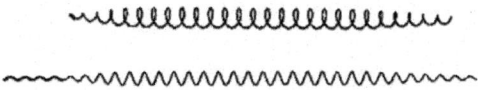

Figure 22: The helical line structure of the light.

The positive magneton on the atomic nucleus à The negative magneton on the electron.

The part of negative magneton on the electron à The radiation of light, the orbit negative magneton.

The radiation of heat, the radiation of light + the orbit negative magneton with same frequency à The positive magneton on the atomic nucleus.

When the magneton that mass is m transform into the radiation of light of half periodic, the velocity of magneton become the velocity of light c, the released kinetic energy is $\dfrac{1}{2}mc^2$.

The light can separate and form again. The diffraction is that the part of light helix is obstructed. The diameter of the light helix is less a little than aperture, the light passing through immediacy if it is at the center of the aperture; the part is obstructed if the light helix is not at the center of the aperture, and change in the directions. The diffraction of twain gap is that the separate light forms again if these phase is same.

The light form again inside the object can explain the color combination. If the diameter of a light helix is less a several fold than another light, a light helix enter into the another light helix inside, the frequency of the combination light is add up, this is the combination of attraction; if the diameter of a light helix is less a little than another light, the frequency of the combination light is reduced, this is the combination of repulsion.

THE MAXIMAL SHELL NUMBERS OF THE ELECTRONIC SHELLS

The negative magnetons on the electron have twenty; the positive magnetons on the atomic nucleus have twenty also, these positive magnetons release out and become the orbit magnetons in the electron orbit, the positive and negative orbit magnetons change to positive magnetons on the nucleus via action; in this cycle, the atomic nucleus release twenty positive magnetons each period. Because the static attraction, the electron will fall on the nucleus if have not the supplement of the positive magnetons, the nucleus change to other nucleus by the action of the electron and nucleus. The number of the positive magnetons that the nucleus released has decided the maximal shell number of the electronic shells, twenty positive magnetons can support the maximal shell number is ten layer, the electronic shells if the shell number exceed ten layer, the electron fall easily on the nucleus, the nucleus is very unstable. If the atomic number is greater than 90, the electronic shells begin exceed ten layer, the most outer shell electrons already cannot form a whole frame, the electrons have no longer subdivide layer. If the atomic number is more, the electron fall easily on the nucleus and bring the nucleus change.

CONCLUSION

Inert gas atoms form a cylinder and a regular polyhedron, forming the same shell electrons. The circle tracks of electrons motion are faces of polyhedron. The electronic number is the faces number of the regular polyhedron. The periodic table of elements is arranged based on a series of regular polyhedrons. The electron has a magnetic outer layer and an electrostatic inner layer. The electron periphery has twenty-petal adsorptive substances, forming a big magnetism radius. If two electrons with the same velocity and direction converge, their adsorptive substances have the same rotational frequency and phase. The magnetic force is $\sqrt{2}$ times the electrostatic force; the existence of adsorptive substance causes the magnetic force greater than the electrostatic force. The magnetic properties of iron, cobalt, and nickel can be explained through the regular dodecahedron electronic shell. The electron orbit converge from reverse direction can explain diamond. The behavior of the adsorptive substances is evolved into the magneton theory, and it can work well with thermodynamic. The free electrons do not have collision in high critical temperature superconductivity. In glass, the same ions will attract each other and can separately form an ion group of a similar type. The structure of the light is a helical line. The maximal shell numbers of the electronic shells are ten.

REFERENCES

1. G. N. Lewis, Journal of the American Chemical Society, Vol. 38, 1916, pp. 762-785.http://dx.doi.org/10.1021/ja02261a002

2. X. T. Lu, H. J. Wang and B. J. Yang, "Atomic Nucleus Physics," Atomic Energy Press, Beijing, 1981, pp. 46-47.

3. O. Echt, K. Sattler and E. Recknagel, Physical Review Letters, Vol. 47, 1981, p. 1121.http://dx.doi.org/10.1103/PhysRevLett.47.1121

4. W. Miehle, O. Kandler, T. Leisner and O. J. Echt, Chemical Physics, Vol. 91, 1989, p. 5940

5. P. W. Stephens and J. G. King, Physical Review Letters, Vol. 51, 1983, pp. 1538-1541.http://dx.doi.org/10.1103/PhysRevLett.51.1538

6. G. S. Tian, "The Origin of Metallic Ferromagnetism," In: 10000 Selected Problems in Sciences, Physics, Science Press, Beijing, 2009, pp. 73-80.

7. H. Stöcker, "Taschenbuch der Physik," Verlag Harri, Deutsch, 1996.

8. A. Chaiken, "Resistivity and specific Heat in Zero Applied Magnetic Field for a Typical Superconductor," 2005. http://en.wikipedia.org/wiki/File:Cvandrhovst.png

9. Y. I. Frenkel, "Kinetic Theory of Liquids," Oxford University Press, Oxford, 1946.

10. M. I. Ojovan, K. P. Travis and R. J. Hand, Journal of Physics: Condensed Matter, Vol. 19, 2007, Article ID: 415107. http://dx.doi.org/10.1088/0953-8984/19/41/415107

11. R. H. Doremus, Journal of Applied Physics, Vol. 92, 2002, pp. 7619-7629.http://dx.doi.org/10.1063/1.1515132

12. M. B. Volf, "Mathematical Approach to Glass," Elsevier, Amsterdam, 1988.

13. M. I. Ojovan and W. E. Lee, Journal of Applied Physics, Vol. 95, 2004, pp. 3803-3810.http://dx.doi.org/10.1063/1.1647260

Chapter 7

MOLECULAR BEAM DEPLETION: A NEW APPROACH

Manuel Dorado
CIRTA, Rotation and Torque Research Center, Theoretical Group, Madrid, Spain

ABSTRACT

During the last years some interesting experimental results have been reported for experiments in N_2O, NO, NO dimer, H_2, Toluene and $BaFCH_3$ cluster. The main result consists in the observation of molecular beam depletion when the molecules of a pulsed beam interact with a static electric or magnetic field and an oscillating field (RF). In these cases, and as a main difference, instead of using four fields as in the original technique developed by I.I. Rabi and others, only two fields, those which configure the resonant unit, are used. That is, without using the nonhomogeneous magnetic fields. The depletion explanation for I.I. Rabi and others is based in the interaction between the molecular electric or magnetic dipole moment and the non-homogeneous fields. But, obviously, the change in the molecules trajectories observed on these new experiments has to be explained without considering the force provided by the field gradient because it happens without using non-homogeneous fields. In this paper a theoretical way for the explanation of these new experimental results is presented. One important point emerges as a result of this development, namely, the existence of an, until now unknown, spin-dependent force which would be responsible of the aforementioned deviation of the molecules.

INTRODUCTION

The molecular beam magnetic resonant (MBMR) technique has significantly contributed, as is well known, to the development of atomic and molecular physics (1). And it makes possible to measure de Larmor frequency of an atom or molecule in the presence of a magnetic field. In the original technique, developed by I.I. Rabi and others [1] -[3] the molecular beam is forced to pass through four different fields:

- A non-homogeneous polarizer field (A) where the molecules are prepared.
- resonant unit (C) that consists of two, a static and an oscillating, fields.
- A non-homogeneous analyzer field (B). Only molecules in the prepared state reach the detector.
- The two non-homogeneous magnetic fields A and B have opposite directions.

The molecular beam describes a sigmoidal trajectory and, finally, is collected in a detector (seeFigure 1).

Rabi explained this effect in terms of spatial reorientation of the angular moment due to a change of state when the transition occurs.

In this case the depletion explanation is based in the interaction between the molecular magnetic dipole moment and the non-homogeneous fields.

$$F = \nabla\left(\boldsymbol{\mu} \cdot \boldsymbol{B}\right) \qquad (1)$$

The force is provided by the field gradient interacting with the molecular dipolar moment (electric or magnetic). On the resonant unit the molecular dipole interact with both, homogeneous and oscillating, fields. When the oscillating field is tuned to a transition resonant frequency between two sub states, a fraction of the molecular beam molecules is removed from the initial prepared state. As a consequence, the dipolar moment changes as well as the interaction force with the nonhomogeneous analyzer field (B). As only molecules in the initial prepared state reach the detector the signal in the detector diminishes.

NEW EXPERIMENTAL RESULTS

During the last years some interesting experimental results have been reported for N_2O, NO, NO dimer, H_2 and $BaFCH_3$ cluster [4] -[7] . The main result consists in the observation of molecular beam depletion when the molecules of a pulsed beam interact with a static electric or magnetic field and an oscillating field (RF) as in the Rabi's experiments. But, in these cases, instead of using four fields, only two fields those which configure the resonant unit (C), are used, that is, without using the non-homogeneous magnetic, A and B, fields. See Figure 2.

In a similar way, when the oscillating field is tuned to a transition resonant frequency between two sub states, the fraction of the molecular beam that is removed from the initial prepared state does not reach the detector.

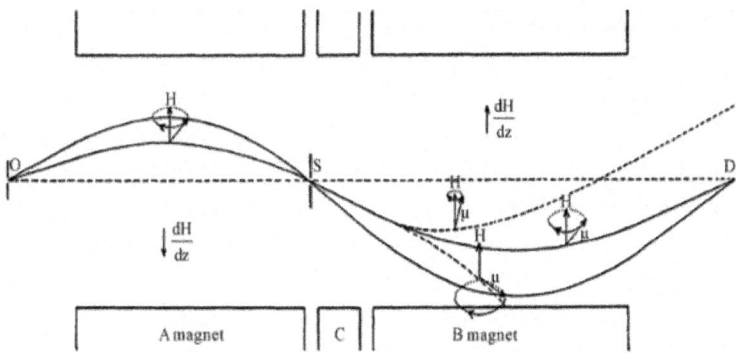

Figure 1: Typical path of molecules in a M.B.M.R experiment. The two solid curves show the paths of the molecules whose moments do not change when passing through the resonant cell.

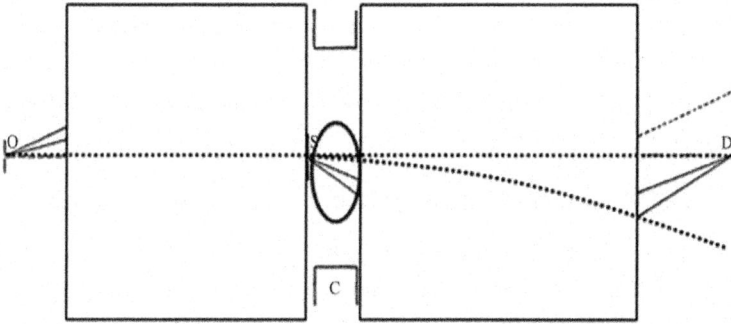

Figure 2: The dotted line path shows the trajectory change of the fraction of the molecular beam that is removed from the initial prepared state when passing thought the resonant cell.

But the important thing is: differently to the previous method, it happens without using non-homogeneous fields. Obviously, the trajectory change has to be explained without considering the force provided by the field gradient. There must be another molecular feature that explains the depletion. It looks as though the linear momentum conservation principle were not satisfied. These experiments suggest that a force depending on other fundamental magnitude of the particle, different from mass and charge must be taken into account.

LOOKING FOR AN EXPLANATION

In order to find out an explanation, let's consider the following case:

An electron is moving, with speed v constant in modulus, in a homogeneous magnetic field B where v is perpendicular to B.

Its kinetic energy will be:

$$E = \frac{1}{2}mv^2$$

(2)

The electron, as is well known, describes a circular trajectory (in general case a helix) with a radius r, being:

$$r = \frac{v}{\omega}$$

and

$$\omega = \frac{qB}{m}$$

(3)

due to the Lorentz force:

$$F = qv \times B$$

(4)

On the other hand, as the electron has a magnetic moment, μ, and spin, S, the presence of the magnetic field B produces a torque when interacting with the electron magnetic moment μ. The angle between S and O_z (the direction of the magnetic field B) remains constant but the spin S revolves about O_z with angular velocity Ω. This phenomenon bears the name of Larmor precession.

The electron kinetic energy must increase with the energy due to spin precession. But it should be considered that the forces producing the torque are perpendicular to the precession motion and, as a consequence, do not modify the energy of the system. It looks like if the principle of energy conservation be violated.

How to Solve This Dilemma?

First Option

If the rotation around an axis is considered as origin of the spin, in a classic (and impossible) interpretation, one could imagine the electron rotating in a slowly way and offsetting the increase in energy due to the precession movement.

But, as it is well known, the spin is a quantized quantity; its modulus is constant and immutable. This option is, as a consequence, not acceptable.

Second Option

Let us consider now if, in our case, helicity is a constant of motion. Helicity, ξ , is defined as the scalar product of linear momentum and the spin:

$$\xi = (mv) \cdot S \tag{5}$$

where S is understood as the classic equivalent of the quantum spin and, as a consequence, $|S|$ remains constant. Is this hypothesis consistent with Quantum Mechanics? Let us consider an electron in a uniform magnetic field B, and let us choose the O_z axis along B. The classical potential energy due to electron magnetic moment μ is then:

$$W = -\mu \cdot B = -\mu_z B = -\gamma B S_z \tag{6}$$

where B is the modulus of the magnetic field. Let us set:

$$W = -\mu \cdot B = -\mu_z B = -\gamma B S_z \tag{7}$$

Ω being the classical angular precession velocity. As is well known, ω_0 has dimensions of the inverse of a time, that is, of an angular velocity. If we replace S_z by the operator S_z the classic energy becomes an operator; the Hamiltonian H which describes the evolution of the spin of the electron in the field B is:

$$H = \omega_0 S_z \tag{8}$$

Since this operator is time independent, solving the corresponding Schrödinger equation amounts to solving the eigenvalue equation of H. We immediately see that the eigenvectors of H are those of S_z (see [8] [9]):

$$H|+\rangle = +\frac{\hbar\omega_0}{2}|+\rangle \tag{9}$$

$$H|-\rangle = -\frac{\hbar\omega_0}{2}|-\rangle \tag{10}$$

There are therefore two energy levels, $E_+ = +\frac{\hbar\omega_0}{2}$ and $E_- = -\frac{\hbar\omega_0}{2}$ Their separation $\hbar\omega_0$ is proportional to the magnetic field and define a single "Bohr frequency"

$$v_{+-} = \frac{1}{h}(E_+ - E_-) = \frac{\omega_0}{2\pi} \tag{11}$$

Is it possible to distinguish, in a uniform magnetic field B, which electrons are the state $|+\rangle$ and which are the state $|-\rangle$? The answer is no. Their behavior inside the field is exactly the same.

But, nevertheless, if we introduce an oscillating magnetic field H_1 with a frequency resonant with the transition $v_{+-} = \frac{1}{h}(E_+ - E_-) = \frac{\omega_0}{2\pi}$, then it will be possible to distinguish both states by the difference in their trajectories, see [10].

LARMOR PRECESSION

We will find, in Quantum Mechanics, the phenomenon equivalent to that described for a particle with classic magnetic moment and spin when moving in a uniform magnetic field B and which bears the name of Larmor precession.

Let us assume that, at time $t = 0$, the spin is in the state

$$|\chi(0)\rangle = \cos\frac{\theta}{2}e^{-i\phi/2}|+\rangle + \sin\frac{\theta}{2}e^{i\phi/2}|-\rangle \tag{12}$$

To calculate the state $|\chi(t)\rangle$ in an arbitrary state $t > 0$ and as $|\chi(0)\rangle$ is already expanded in terms of the eigenstates of the Hamiltonian we will obtain

$$|\chi(t)\rangle = \cos\frac{\theta}{2}e^{-\frac{i\phi}{2}}e^{-iE_+\frac{t}{h}}|+\rangle + \sin\frac{\theta}{2}e^{li\phi/2e^{-iE_-\frac{t}{h}}}|-\rangle \tag{13}$$

Or, using the values of E_+ and E_-:

$$|\chi(t)\rangle = \cos\frac{\theta}{2}e^{-\frac{i(\phi+\omega_0 t)}{2}}|+\rangle + \sin\frac{\theta}{2}e^{+\frac{i(\phi+\omega_0 t)}{2}}|-\rangle \tag{14}$$

The presence of the magnetic field B therefore introduces a phase shift, proportional to time, between the coefficients of the kets $|+\rangle$ and $|-\rangle$.

Comparing the equation (14) for $|\chi(t)\rangle$ with that for the eingenket $|+\rangle_u$ for the observable $S \cdot u$

$$|+\rangle_u = \cos\frac{\theta}{2}e^{-i\phi/2}|+\rangle + \sin\frac{\theta}{2}e^{+\frac{i\phi}{2}}|-\rangle \tag{15}$$

We see that the direction $u(t)$ along which the component is, with certainty, is defined by the polar angles:

$$\begin{cases} \theta_t = \theta = Cte. \\ \phi(t) = \phi + \omega_0 t \end{cases}$$

(16)

that coincides with the direction along which the classic spin should be pointing out.

The angle between $u(t)$ and O_z (the direction of the magnetic field B) therefore remains constant, but $u(t)$ revolves around O_z with angular velocity $\omega_0 = \Omega$ proportional to the magnetic field. And the mean values of S_x, S_y and S_z behave like the components of a classical angular momentum of modulus constant $\frac{1}{2}\hbar$ undergoing Larmor precession.

Helicity as a Constant of Motion

As is well known, helicity is not, in general, a constant of motion. The reason is that helicity operator does not commute, in general, with the Hamiltonian H. Nevertheless, it will be proven that, at least for the interaction here considered (Larmor), the helicity eigenvalue is conserved along the electron's (classical) trajectory.

We redefine now the helicity, ξ, in order that its eigenvalues be ± 1, as $\xi = \sigma \hat{v}$, where $\hat{v} = \dfrac{v}{v}$ and $s = \dfrac{\hbar}{2}\sigma$ .The initial velocity of the electron is $v = v_0 (\cos\phi, \sin\phi, 0)$, and we assume the initial spin state of the electron to be an eigenstate of the helicity with eigenvalue +1, which is given in equation (12), with $\theta = \dfrac{\pi}{2}$ that is:

$$|\chi(0)\rangle = \frac{\sqrt{2}}{2}\left(e^{-\frac{i\phi}{2}}|+\rangle + e^{\frac{i\phi}{2}}|-\rangle \right)$$

(17)

At the time t the velocity of the electron is, as it is known,

$$v(t) = v_0\left[\cos(\phi + \omega_0 t), \sin(\phi + \omega_0 t), 0 \right]$$

(18)

where ω_0 is given in Equation (7). According to Equations (15) and (16), with $\theta = \frac{\pi}{2}$, at time t the spin state is:

$$\left|\chi(t)\right\rangle = \frac{\sqrt{2}}{2}\left(e^{-\frac{i(\phi+\omega_0 t)}{2}}\left|+\right\rangle + e^{+\frac{i(\phi+\omega_0 t)}{2}}\left|-\right\rangle\right)$$

(19)

and the helicity at time t, $\xi(t) = \sigma\hat{v}(t)$, , ,

$$\xi(t) = \sigma_x \cos(\phi + \omega_0 t) + \sigma_y \sin(\phi + \omega_0 t)$$

(20)

Now, taking into account that,

$$\sigma_x\left(\left|+\right\rangle, \left|-\right\rangle\right) = \left(\left|-\right\rangle, \left|+\right\rangle\right); \sigma_y\left(\left|+\right\rangle, \left|-\right\rangle\right) = \left(i\left|-\right\rangle, -i\left|+\right\rangle\right);$$

We easily obtain:

$$\xi(t)\left|\chi(t)\right\rangle = \left|\chi(t)\right\rangle$$

(21)

This shows that $\left|\chi(t)\right\rangle$ is an eigenstate of the helicity of eigenvalue +1; in other words, helicity is conserved along the electron's (classical) trajectory.

CONSEQUENCES

It has been proven, for cases here considered, that helicity is a constant of motion. As a consequence of this result, the linear momentum mv must have the same precession angular velocity (Larmor angular velocity) Ω than the spin S. The equation of motion describing the linear momentum evolution must be then equivalent of the equation of motion which describe the evolution of the spin S. This means that:

$$\left(\frac{dS}{dt}\right)_{\text{Inertial}} = S \times \Omega$$

(22)

$$\left(\frac{dmv}{dt}\right)_{\text{Inertial}} = mv \times \Omega$$

(23)

It is concluded the particle will be under a central acceleration, $a = v \times \Omega$ perpendicular to v. The particle is then under a central force:

$$F = mv \times \Omega$$

(24)

This kind of forces related with the spin will be designed as Lorentz-like forces. In this case, the trajectory will be a circular one. The radius will be:

$$R = \frac{|v|}{|\Omega|}$$

(25)

And its kinetic energy:

$$E = \frac{1}{2}I_z\Omega^2 = \frac{1}{2}mR^2\Omega^2 = \frac{1}{2}mv^2$$

(26)

which is equal to the initial one shown in Equation (2). The force (Equation (24)) is the responsible of the electron circular trajectory inside the field B and should be related to the spin S of the electron.

In conclusion, as helicity is, in those cases, a constant of motion, the particle is under a Lorentz-like force and the principles of conservation are not violated.

Electron in a Magnetic Field

If the case of an electron in a magnetic field is considered, then the force due to the spin of the electron will be:

$$F = mv \times \Omega$$

where Ω is the spin Larmor precession velocity around O_z. But is known that:

$$\Omega = \frac{|\mu|B}{|S|} = \frac{gqB}{2m}$$

(27)

Substituting in Equation (24) the expression for the force acting on the particle is obtained. This force has its origin on the spin. This expression is:

$$F = \frac{mg}{2m}qv \times B$$

(28)

As for an electron $g = 2$, the final result is:

$$F = qv \times B$$

(29)

Surprisingly this expression for the Lorentz-like force related to the spin coincides, formally, with that known as Lorentz force related to the charge. Considering the spin as responsible of the Lorentz-like force, a new deflection mechanism has been proposed (see [10]). The equations of motion for a system with intrinsic angular momentum when applying torques are described and, according with the theory, when the frequency of the oscillating field coincides with a transition resonant frequency (Larmor frequency), the molecules that change their state from the original one are removed from their trajectories and, as a consequence, do not reach the detector and the corresponding signal decreases.

NEW EXPERIMENTAL PROPOSAL

In 1939 Alvarez and Bloch [11] measured the neutron magnetic moment by using a neutron beam passing through a resonant unit. Neutrons from the Be + D reactions were slowed to thermal velocities and diffused down a Cadmium lined tube through the water tank to the polarizer magnet, B_A. After passing through the resonant unit that consists of two, a static and an oscillating, fields and the analyser magnet, B_B they were detected in a BF_3 chamber. The polarizer B_A and analyser B_B are strongly magnetized iron pieces. A neutron resonant dip is observed in the signal of the neutron beam when the oscillating resonant frequency corresponding to the transition between the two states up and down is reached. According to the previous theoretical description and recently results obtained for NO_2, NO, NO dimer, H_2 and $BaFCH_3$ cluster, if the Álvarez and Bloch experiment is carried out without using analyser magnet B_B, we anticipate that the experimental results will be the same as those obtained by Álvarez and Bloch in the experiment of 1939.

According to the new explanation, the trajectory change takes place when neutrons pass through the resonant unit and the oscillating field is tuned to a transition resonant frequency between two states, up and down, of the spinof the neutron. In case of Álvarez and Bloch experiment, they used a magnetic field for the neutron resonance of 622 Gauss and a resonant frequency of oscillator of 1843 kilocycles.

SUMMARIZING

- Some recent experimental results have reported the observation of molecular beam depletion when molecules of a pulse interact with a homogeneous static electric or magnetic field and an oscillating field (RF).

- In absence of non-homogeneous fields it is not possible to use the force provided by the field gradient interaction with the molecular dipole in order to explain this depletion.

- A unknown force depending on other fundamental magnitude of the particle, different of mass and charge must be considered.

CONCLUSIONS

To the best of our knowledge, it seems that existence of forces described in this paper, related with the spin of the particles, is the more adequate way to explain, from a theoretical point of view, the experimental result here considered. These forces are called Lorentz-like forces.

However, more experimental works are needed to support this conclusion. In this sense the experiment with neutrons, suggested in our proposal, is a very good example of a relevant experiment to be carried out.

ACKNOWLEDGEMENTS

The author is very grateful to prof. José L. Sánchez Gómez, Universidad Autónoma de Madrid and to Prof. Ramón Fernández Álvarez-Estrada, Universidad Complutense de Madrid, for useful discussions.

REFERENCES

1. Ramsey, N.F. (1990) Molecular Beams. Oxford University Press, Oxford.

2. Rabi, I.I. (1935) Physical Review, 49, 324. http://dx.doi.org/10.1103/PhysRev.49.324

3. Rabi, I.I., Millman, S., Kusch, P. and Zacharias, J.R. (1939) Physical Review, 55, 526.http://dx.doi.org/10.1103/PhysRev.55.526

4. Gonzalez Urena, A., Gamo, L., Gasmi, K., Caceres, J.O., De Castro, M., Skowronek, S., Dorado, M., Morales Furio, M., Perez, J.L. and Sanchez Gomez, J.L. (2001) Chemical Physics Letters, 341, 495.http://dx.doi.org/10.1016/S0009-2614(01)00529-2

5. Montero, C., Gonzalez Urena, A. and Morato, M. (2003) European Journal of Physics D, 26, 261-264.http://dx.doi.org/10.1140/epjd/e2003-00259-5

6. Gasmi, K., Gonzalvez, A.G. and Gonzalez Urena, A. (2010) Journal of Physical Chemistry A, 114, 3229-3236. http://dx.doi.org/10.1021/jp909398w

7. Montero, C., Garcia Urena, J. and Dorado, M. (2014) New Transition Mechanism: The Shaking Effect. http://arxiv.org/abs/1402.0203

8. Dorado Gonzalez, M. (1997) Anales de Fisica, 93, 105-120. arXiv:physics/0101085

9. Cohen Tannoudji, C., Diu, B. and Laloe, F. (2005) Quantum Mechanics. Vols. I and II. Wiley-VCH, Hoboken.

10. Dorado Gonzalez, M. (2013) Dinamica de sistemas con espin: Un nuevo enfoque. Fundamentos y aplicaciones (in Spanish). ADI Servicios Editoriales.

11. Alvarez, L.W. and Bloch, F. (1940) Physical Review, 57, 111.http://dx.doi.org/10.1103/PhysRev.57.111

Chapter 8

DUAL CONJUGATE ADAPTIVE OPTICS PROTOTYPE FOR WIDE FIELD HIGH RESOLUTION RETINAL IMAGING

Zoran Popovic[1], Jörgen Thaung[1], Per Knutsson[1] and Mette Owner-Petersen[2]

[1] Department of Ophthalmology, University of Gothenburg, Gothenburg, Sweden

[2] Retired from the Telescope Group, Lund University, Lund, Sweden

INTRODUCTION

Retinal imaging is limited due to optical aberrations caused by imperfections in the optical media of the eye. Consequently, diffraction limited retinal imaging can be achieved if optical aberrations in the eye are measured and corrected. Information about retinal pathology and structure on a cellular level is thus not available in a clinical setting but only from histological studies of excised retinal tissue. In addition to limitations such as tissue shrinkage and distortion, the main limitation of histological preparations is that longitudinal studies of disease progression and/or results of medical treatment are not possible.

Adaptive optics (AO) is the science, technology and art of capturing diffraction-limited images in adverse circumstances that would normally lead to strongly degraded image quality and loss of resolution. In non-military applications, it was first proposed and implemented in astronomy [1]. AO technology has since been applied in many disciplines, including vision science, where retinal features down to a few microns can be resolved by correcting the aberrations of ocular optics. As the focus of this chapter is on AO retinal imaging, we will focus our description to this particular field.

The general principle of AO is to measure the aberrations introduced by the media between an object of interest and its image with a wavefront sensor, analyze the measurements, and calculate a correction with a control computer.

The corrections are applied to a deformable mirror (DM) positioned in the optical path between the object and its image, thereby enabling high-resolution imaging of the object.

Modern telescopes with integrated AO systems employ the laser guide star technique [2] to create an artificial reference object above the earth's atmosphere. Analogously, the vast majority of present-day vision research AO systems employ a single point source on the retina as a reference object for aberration measurements, consequently termed guide star (GS). AO correction is accomplished with a single DM in a plane conjugated to the pupil plane. An AO system with one GS and one DM will henceforth be referred to as single-conjugate AO (SCAO) system. Aberrations in such a system are measured for a single field angle and correction is uniformly applied over the entire field of view (FOV). Since the eye's optical aberrations are dependent on the field angle this will result in a small corrected FOV of approximately 2 degrees [3]. The property of non-uniformity is shared by most optical aberrations such as e.g. the well known primary aberrations of coma, astigmatism, field curvature and distortion.

A method to deal with this limitation of SCAO was first proposed by Dicke [4] and later developed by Beckers [5]. The proposed method is known as multiconjugate AO (MCAO) and uses multiple DMs conjugated to separate turbulent layers of the atmosphere and several GS to increase the corrected FOV. In theory, correcting (in reverse order) for each turbulent layer could yield diffraction limited performance over the entire FOV. However, as is the case for both the atmosphere and the eye, aberrations do not originate solely from a discrete set of thin layers but from a distributed volume. By measuring aberrations in different angular directions using several GSs and correcting aberrations in several layers of the eye using multiple DMs (at least two), it is possible to correct aberrations over a larger FOV than compared to SCAO.

The concept of MCAO for astronomy has been the studied extensively [6-12], a number of experimental papers have also been published [13-16], and on-sky experiments have recently been launched [17]. However, MCAO for the eye is just emerging, with only a few published theoretical papers [3, 18-21]. Our group recently published the first experimental study [21] and practical application [22] of this technique in the eye, implementing a laboratory demonstrator comprising multiple GSs and two DMs, consequently termed dual-conjugate adaptive optics (DCAO). It enables imaging of retinal features down to a few microns, such as retinal cone photoreceptors and capillaries [22], the smallest blood vessels in the retina, over an imaging area of approximately 7 x 7 deg^2. It is unique in its ability to acquire single images over a retinal area that is up to 50 times larger than most other research based flood illumination AO instruments, thus potentially allowing for clinical use.

A second-generation Proof-of-Concept (PoC) prototype based on the DCAO laboratory demonstrator is currently under construction and features several improvements. Most significant among those are changing the order in which DM corrections are imposed and the implementation of a novel concept for multiple GS creation (patent pending).

BRIEF ANATOMICAL DESCRIPTION OF THE EYE

The human eye can be divided into an optical part and a sensory part. Much like a photographic lens relays light to an image plane in a camera, the optics of the eye consisting of the cornea, the pupil, and the lens, project light from the outside world to the sensory retina (Fig. 1, left). The amount of light that enters the eye is controlled by pupil constriction and dilation. The human retina is a layered structure approximately 250 µm thick [23, 24], with a variety of neurons arranged in layers and interconnected with synapses (Fig. 1, right).

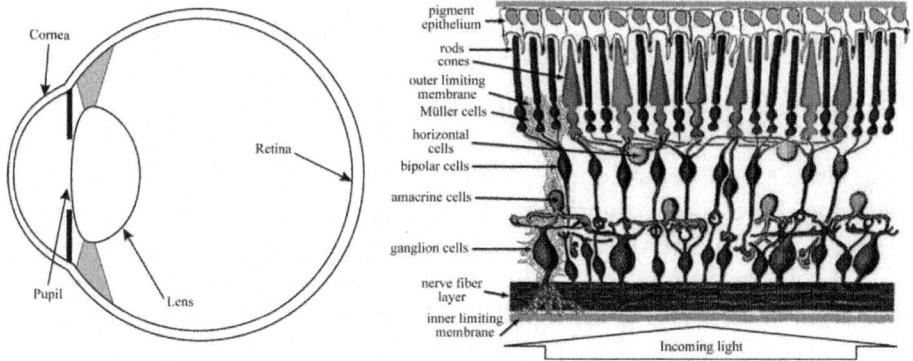

Figure 1: Schematic drawings of the eye (left) and the layered retinal structure (right). (Webvision, http://webvision.med.utah.edu/book/part-i-foundations/simple-anatomy-of-the-retina/).

Visual input is transformed in the retina to electrical signals that are transmitted via the optic nerve to the visual cortex in the brain. This process begins with the absorption of photons in the retinal photoreceptors, situated at the back of the retina, which stimulate several interneurons that in turn relay signals to the output neurons, the retinal ganglion cells. The ganglion cell nerve fiber axons exit the eye through the optic nerve head (blind spot).

Unlike the regularly spaced pixels of equal size in a CCD chip the retinal photoreceptor mosaic is an inhomogeneous distribution of cone and rod photoreceptors of various sizes. The central retina is cone-dominated with a cone density peak at the fovea, the most central part of the retina responsible for sharp vision, with a decrease in density towards the rod-dominated periphery. Cones are used for color and photopic (day) vision and rods are used for scotopic (night) vision.

Blood is supplied to the retina through the choroidal and retinal blood vessels. The choroidal vessels line the outside of the eye and supply nourishment to the photoreceptors and outer retina, while the retinal vessels supply inner retinal layers with blood. Retinal capillaries, the smallest blood vessels in the eye, branch off from retinal arteries to form an intricate network throughout the whole retina with the exception of the foveal avascular zone (FAZ). The FAZ is the capillary-free region of the fovea that contains the foveal pit where the cones are most densely packed and are completely exposed to incoming light. Capillaries form a superficial layer in the nerve fiber layer, a second layer in the ganglion cell layer, and a third layer running deeper into the retina.

BRIEF THEORETICAL BACKGROUND

AO CALIBRATION PROCEDURE

The AO concept requires a procedure for calculating actuator commands based on WFS signals relative to a defined set of zero points, so-called calibration. Both the DCAO demonstrator and the PoC prototype are calibrated using the same direct slope algorithm. The purpose is to construct an interaction matrix G by calculating the sensor response $s = [s_1, s_2,..., s_m]^T$ to a sequence of DM actuator commands $c = [c_1, c_2,..., c_n]^T$. Here s is a vector of measured wavefront slopes, m/2 is the number of subapertures, and n is the number of DM actuators. This relation is defined by

$$S = Gc \tag{1}$$

and the interaction matrix is given by

$$G = \begin{bmatrix} \partial s_1/\partial c_1 & \partial s_1/\partial c_2 & \cdots & \partial s_1/\partial c_n \\ \partial s_2/\partial c_1 & \partial s_2/\partial c_2 & \cdots & \partial s_2/\partial c_n \\ \vdots & \vdots & & \vdots \\ \partial s_m/\partial c_1 & \partial s_m/\partial c_2 & \cdots & \partial s_m/\partial c_n \end{bmatrix}. \tag{2}$$

The relation above has to be modified to allow for multiple GSs and DMs by concatenating multiple s and c vectors. In the case of five GSs and two DMs we obtain

$$\begin{pmatrix} s_1 \\ s_2 \\ s_3 \\ s_4 \\ s_5 \end{pmatrix} = G \begin{pmatrix} c_1 \\ c_2 \end{pmatrix},$$

(3)

Where

$$G = \begin{bmatrix} \partial s_1/\partial c_1 & \partial s_1/\partial c_2 \\ \partial s_2/\partial c_1 & \partial s_2/\partial c_2 \\ \partial s_3/\partial c_1 & \partial s_3/\partial c_2 \\ \partial s_4/\partial c_1 & \partial s_4/\partial c_2 \\ \partial s_5/\partial c_1 & \partial s_5/\partial c_2 \end{bmatrix}.$$

(4)

The interaction matrix G is constructed by poking each DM actuator in sequence with a positive and a negative unit poke and calculating an average response, starting with the first actuator on DM1 and ending with the last actuator on DM2. In the case of five Hartmann patterns with 129 subapertures each and two DMs with a total of 149 actuators we obtain an interaction matrix dimension of 1290×149. The reconstructor matrix G^+ is calculated using singular value decomposition (SVD) [25] since

$$G = U \Lambda V^T,$$

(5)

where U is an m×m unitary matrix, Λ is an m×n diagonal matrix with nonzero diagonal elements and all other elements equal to zero, and V^T is the transpose of V, an n×n unitary matrix. The non-zero diagonal elements λi of Λ are the singular values of G. The pseudoinverse of G can now be computed as

$$G^+ = V \Lambda^+ U^T,$$

(6)

which is also the least squares solution to Eq. (1). The diagonal values of Λ^+ are set to λi^{-1}, or zero if λi is less than a defined threshold value. Non-zero singular values correspond to correctable modes of the system. Noise sensitivity can be reduced by removing modes with very small singular values. DM actuator commands can then be calculated by matrix multiplication:

$$\begin{pmatrix} c_1 \\ c_2 \end{pmatrix} = G^+ \begin{pmatrix} s_1 \\ s_2 \\ s_3 \\ s_4 \\ s_5 \end{pmatrix}.$$

(7)

However, even the most meticulous calibration of DM and WFS interaction will not yield optimal imaging performance due to non-common path errors between the wavefront sensor and the final focal plane of the imaging channel. The reduction of these effects by proper zero point calibration is therefore crucial to achieve optimal performance of an AO system. Several methods have been proposed to improve imaging performance [26-33]. The method implemented in our system is similar to the imaging sharpening method [29, 30], but a novel figure of merit is used, and the inherent singular modes of the AO system are optimized (patent pending).

CORRECTED FIELD OF VIEW

In SCAO a single GS is used to measure wavefront aberrations and a single DM is used to correct the aberrations in the pupil plane. This will result in a small corrected FOV due to field dependent aberrations in the eye. However, the corrected FOV in the eye can be increased by using several GS distributed across the FOV and two or more DMs [3, 19-21]. A larger FOV than in SCAO can actually be obtained by using several GS and a single DM in the pupil plane, analogous to ground layer AO (GLAO) in astronomy [34], but the increase in FOV size and the magnitude of correction will be less than when using multiple DMs.

A relative comparison of simulated corrected FOV for the three cases of SCAO, GLAO, and DCAO in our setup is shown in Fig. 2. The simulated FOV is approximately 7×7 degrees, with a centrally positioned GS in the SCAO simulation and five GS positioned in an 'X' formation with the four peripheral GSs displaced from the central GS by a visual angle of 3.1 deg in the GLAO and DCAO simulations.

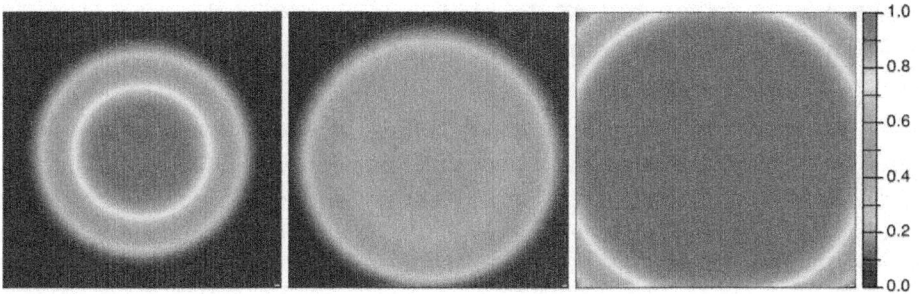

Figure 2: Zemax simulation of a corrected 7×7 deg FOV in our setup using the Liou-Brennan eye model [35] for SCAO (left), multiple GS and single DM (middle), and DCAO (right). Color bar represents simulated Strehl ratio.

EXPERIMENTAL SETUPS

DCAO DEMONSTRATOR

Only a basic description highlighting modifications to the original DCAO demonstrator will be given here. The reader is referred to [21] for a detailed description of the setup. The basic layout of the DCAO demonstrator is shown in Fig. 3.

Dcao Demonstrator Wavefront Measurement And Correction

Continuous, relatively broadband (to avoid speckle effects), near-infrared light (834±13 nm) from a super-luminescent diode (SLD), delivered through a 1:5 fiber splitter and five single mode fibers, is used to generate the five GS beams. The advantage of using an SLD as a source is that the short coherence length of the SLD light generates much less speckle in the Shack-Hartmann WFS spots than a coherent laser source. The end ferrules of the single mode fibers are mounted in a custom fiber holder and create an array of point sources, which are imaged via the DMs and a Badal focus corrector onto the retina. The GSs are arranged in an 'X' formation, with the four peripheral GSs displaced from the central GS by a visual angle of 3.1 deg, corresponding to a retinal separation of approximately 880 μm in an emmetropic eye.

Reflected light from the GSs passes through the optical media of the eye and emerges through the pupil as five aberrated wavefronts. After the Badal focus corrector and the two DMs the light passes through a collimating lens array (CLA) consisting of five identical lenses, one for each GS. The five beams are focused by a lens (L7) to a common focal point (c.f. Fig. 8), collimated by a lens (L8) and individually sampled by the WFS, an arrangement consequently termed multi-reference WFS. In addition to separating the WFS Hartmann patterns as in [36] this arrangement makes it possible to filter light from all five GSs using a single pinhole (US Patent 7,639,369).

Custom written AO software for control of one or two DM and one to five GS was developed, tested, and implemented by Landell [37]. The pupil DM (DM1) will apply an identical correction for all field-points in the FOV. The second DM (DM2), positioned in a plane conjugated to a plane approximately 3 mm in front of the retina, will contribute with partially individual corrections for the five angular directions and thus compensate for non-uniform (anisoplanatic) or field-dependent aberrations. The location of DM2 was chosen to ensure an smooth correction over the FOV by allowing sufficient overlap of GS beam footprints.

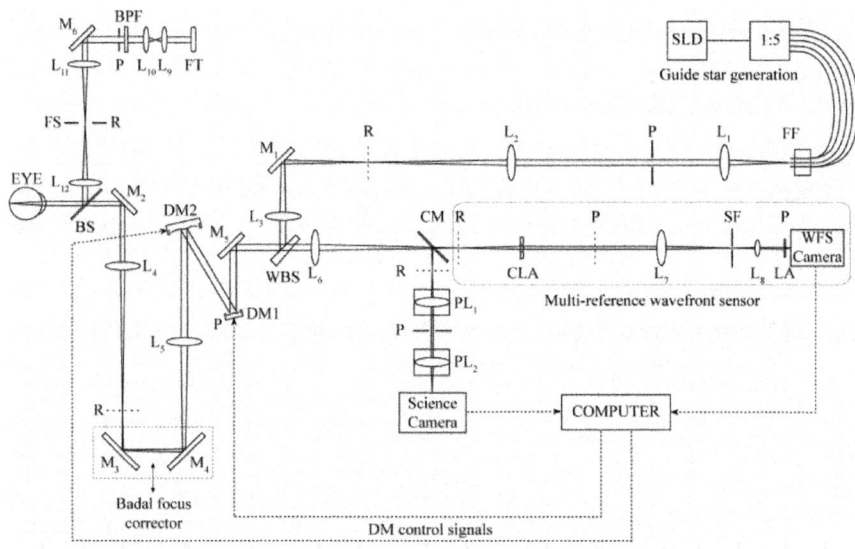

Figure 3: Basic layout of the DCAO demonstrator. Abbreviations: BPF – band-pass filter, BS – beamsplitter, CLA – collimating lens array, CM – cold mirror, DM1 – pupil DM, DM2 – field DM, FF – fiber ferrules, FS – field stop, FT – flash tube, LA – lens-let array, M – mirror, P – pupil conjugate plane, PL – photographic lens, PM – pupil mask, R – retinal conjugate plane, SF – spatial filter, SLD – superluminescent diode, WBS – wedge beamsplitter.

Dcao Demonstrator Retinal Imaging

For imaging purposes, the retina is illuminated with a flash from a Xenon flash lamp, filtered by a 575±10 nm wavelength bandpass filter (BP). The narrow bandwidth of the BP is essential to minimize chromatic errors, in particular longitudinal chromatic aberration (LCA) [38] in the image plane of the retinal camera.

The illuminated field on the retina (approximately 10×10 degrees) is limited by a square field stop in a retinal conjugate plane. Visible light from the eye is reflected by a cold mirror (CM) and relayed through a pair of matched photographic lenses, chosen to minimize non-common path errors. An adjustable iris between the two photographic lenses is used to set the pupil size used for imaging, corresponding to a diameter of 6 mm at the eye.

Imaging is performed with a science grade monochromatic CCD science camera with 2048×2048 pixels and a square pixel cell size of 7.4 μm is used for imaging. The size of the CCD chip corresponds to a retinal FOV of 6.7×6.7 deg². The full width at half maximum (FWHM) of the Airy disk in the image plane at 575 nm is 15 μm and hence the image is sampled according to the Nyquist-Shannon sampling theorem (two pixels per FWHM).

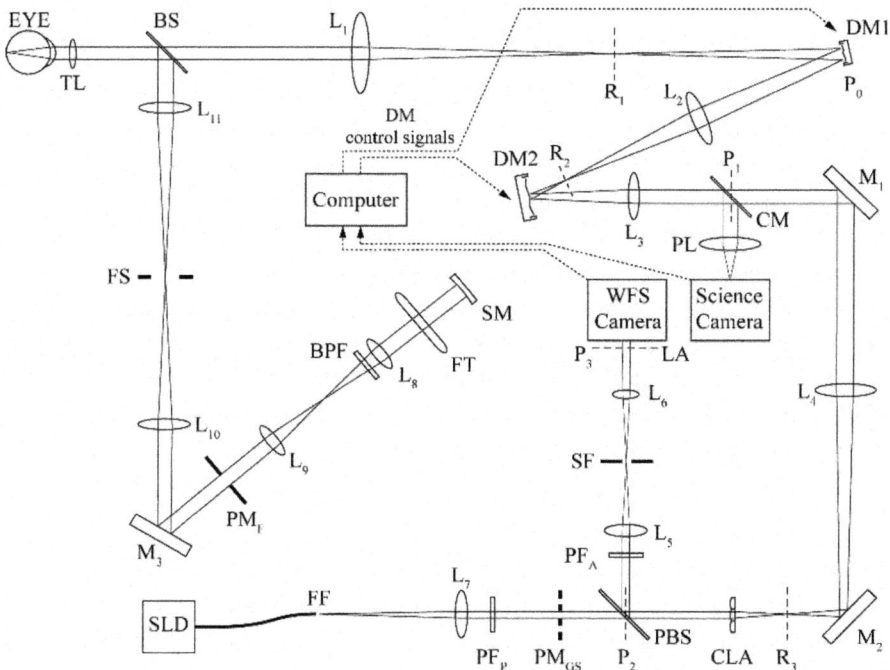

Figure 4: Basic layout of the PoC prototype. Abbreviations: BPF – band-pass filter, BS – beamsplitter, CLA – collimating lens array, CM – cold mirror, DM1 – pupil DM, DM2 – field DM, FS –field stop, FT – flash tube, LA – lenslet array, M – mirror, P – pupil conjugate plane, PBS – pellicle beamsplitter, PF$_P$/PF$_A$ – polarization filters, PL – photographic lens, PM$_F$ – flash pupil mask, PM$_{GS}$ – GS pupil mask, R – retinal

conjugate plane, SF – spatial filter, SM – spherical mirror, SLD – superluminescent diode, TL – trial lens. Fixed corrective lenses are either lens pairs or single lenses.

Poc Prototype

A PoC prototype (Fig. 4) has been developed to evaluate the clinical relevance of DCAO wide-field high-resolution retinal imaging. The prototype is currently under construction and features several improvements with regards to the DCAO demonstrator. Most significant among those are that the order in which DM corrections are imposed has been changed and a novel implementation of GS creation (patent pending). The size of the PoC prototype has been greatly reduced compared with the optical table design of the DCAO demonstrator to a compact joystick operated tabletop instrument 600×170×680 mm (H×W×D) in size. The opto-mechanical layout comprises five modules: a GS generation module, a main module, a WFS module, a flash module, and an imaging module.

Poc Gs Generation Module

A novel method of GS creation has been implemented in the PoC prototype, whereby the CLA that is part of the WFS is also utilized to create the GS beams. Collimated 835 ± 10 nm SLD light from a single mode fiber is polarized (PF_p) and passes through a multi-aperture stop with five apertures (PM_{GS}) that are aligned to the five CLA lenses. Since the CLA is used for GS generation and also enables single point spatial filtering in the multi-reference WFS we have an auto-collimating arrangement that greatly reduces system complexity and alignment. The GS rays pass through standard and custom relay optics and the DMs before entering the eye, where they form five spots arranged in an 'X' formation. The four peripheral GSs are diagonally displaced from the central GS by a visual angle of 3.1 deg (880 μm) on the retina.

Poc Main Module

Residual focus and astigmatism aberrations in the DCAO demonstrator that had not been compensated for by a Badal focus corrector and trial astigmatism lenses were corrected by DM1 after passing DM2, resulting in sub-optimal DM2 performance. The PoC prototype features a correct arrangement of the

DMs where reflected light from the eye, corrected by trial lenses, first passes the pupil mirror DM1 before passing the field mirror DM2.

DM1 is a Hi-Speed DM52-15 (ALPAO S.A.S., Grenoble, France), a 52 actuator magnetic DM with a 9 mm diameter optical surface and 1.5 mm actuator separation. The magnification relative to the pupil of the eye is 1.5, thus setting the effective pupil area of the instrument to 6 mm at the eye. DM2 is a Hi-Speed DM97-15 (ALPAO S.A.S., Grenoble, France), a 97 actuator magnetic DM with a 13.5 mm diameter optical surface and 1.5 mm actuator separation. GS beam footprints on DM1 and DM2 are shown in Fig. 5. The last element of the main module is a dichroic beamsplitter (CM) that reflects collimated imaging light towards the retinal camera and transmits collimated GS light towards the WFS.

As the relay optics of the main module transmits both measurement (835 nm) and imaging (575 nm) light, custom optics were designed to assure diffraction limited performance at both wavelengths (Fig. 6). Due to the ocular chromatic aberrations the bandwidth of the flash illumination bandpass filter will induce a wavelength dependent focal shift in the instrument image plane. An evaluation of the focal shift for the 575±10 nm wavelengths transmitted by the flash illumination bandpass filter using the Liou-Brennan Zemax eye model [35] yields a ±6.9 µm focal shift at the retina (Fig. 7).

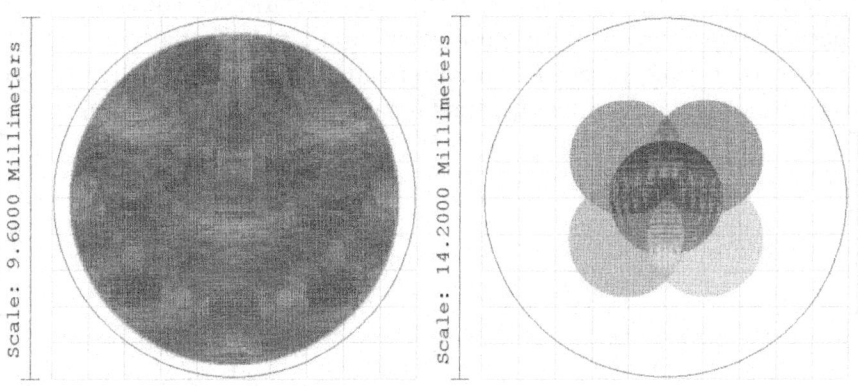

Figure 5: GS beam footprints on DM1 (left) and DM2 (right).

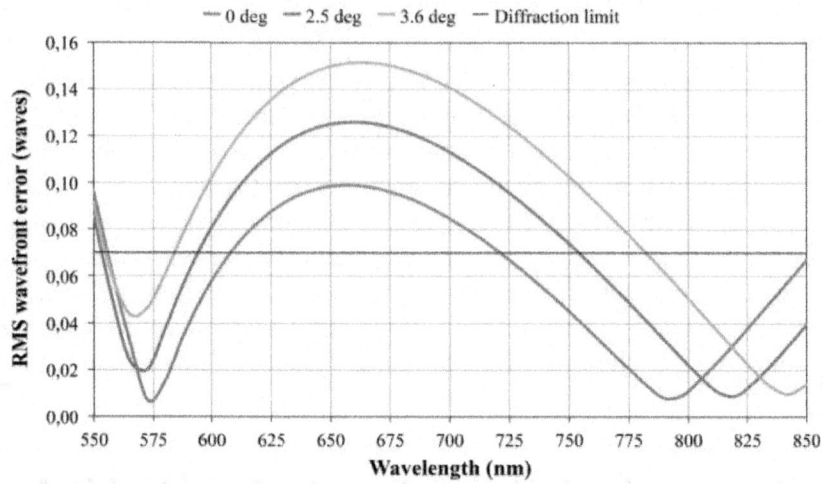

Figure 6: RMS wavefront error of the PoC main module custom relay optics at the main module exit pupil for three retinal field positions (0, 2.5, and 3.6 deg).

Poc Wfs Module

A multi-reference WFS with spatial filtering (Fig. 8) has been implemented in both the DCAO demonstrator and the PoC prototype. The design greatly reduces system complexity by implementing a single spatial filter to reduce unwanted light from parasitic source reflections and scattered light from the retina when imaging multiple Hartmann patterns with a single WFS camera.

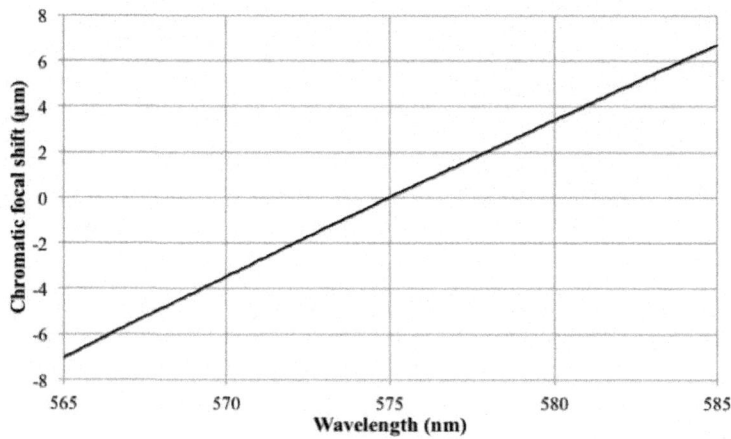

Figure 7: Chromatic focal shift over flash illumination bandpass filter bandwidth (575±10 nm) at the retina calculated using the Liou-Brennan eye model [35].

Transmitted GS light from the main module passes through the CLA and is reflected by a pellicle beam splitter. A second polarizing filter (PF_A) removes unwanted backscattered reflections from the GS generation, and a lens brings the five GS beams to a common focus where they are spatially filtered by a single aperture (SF). A collimating lens finally relays the five beams onto a lenslet array (LA) with a focal length of 3.45 mm and a lenslet pitch of 130 μm.

The monochromatic WFS CCD camera has 1388×1038 pixels with a square pixel cell size of 6.45 μm, of which a central ROI of 964×964 pixels is used for wavefront sensing. The diameter of the diffraction limited focus spot of a lenslet is 2.44 λ f / d = 54 μm. Each spot will consequently be sampled by approximately 8×8 pixels, an oversampling that can be alleviated using pixel binning. The 6 mm pupil diameter of the eye is demagnified to 1.87 mm at the WFS and each Hartmann pattern will consequently be sampled by ~13 lenslets across the diameter (Fig. 9).

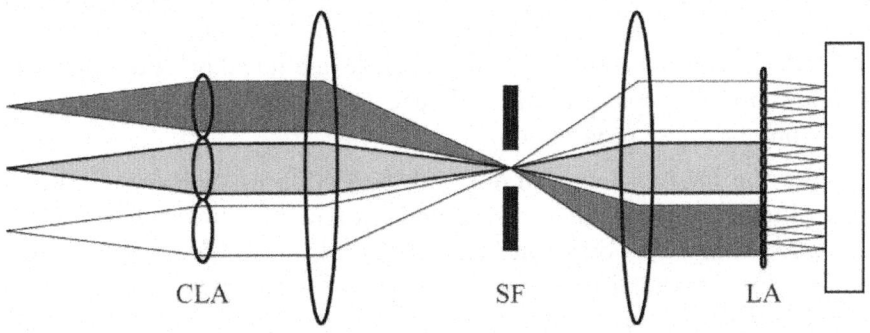

Figure 8: Schematic drawing of the multi-reference WFS with spatial filtering.

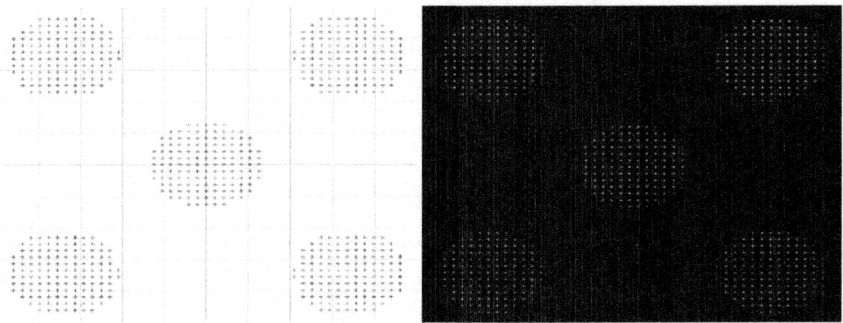

Figure 9: Zemax simulation of Hartmann spot image (left) and actual WFS image (right).

Poc Flash And Imaging Modules

Retinal images are obtained by illuminating a 10×10 degree retinal field using a 4-6 ms spectrally filtered (575±10 nm) Xenon flash. A Canon EF 135mm f/2.0 L photographic lens is used to focus reflected light from the dichroic beamsplitter onto the science camera, a 2452×2056 pixel Stingray F-504B monochromatic CCD with a square pixel cell size of 3.45 μm (Allied Vision Technologies GmbH, Stadtroda, Germany). The physical size of the full chip corresponds to a retinal FOV of 8.28×6.94 deg with a pixel resolution of 0.059 mrad (0.974 μm on the retina).

RETINAL IMAGING

AO retinal imaging reveals information about retinal structures and pathology currently not available in a clinical setting. The resolution of retinal features on a cellular level offers the possibility to reveal microscopic changes during the earliest stages of a retinal disease. One of the most important future applications of this technique is consequently in clinical practice where it will facilitate early diagnosis of retinal disease, follow-up of treatment effects, and follow-up of disease progression.

Both the DCAO demonstrator and the PoC prototype feature a narrow depth of focus, approximately 25 μm and 9 μm in the retina, respectively. This allows for imaging of different retinal layers, from the deeper photoreceptor layer to the superficial blood vessel and nerve fiber layers. Images are flat-fielded using a low-pass filtered image to reduce uneven illumination [39]. A Gaussian kernel with σ = 8 - 25 pixels is chosen depending on the imaged retinal layer. A smaller kernel is used for images of the photoreceptor layer and a larger kernel is consequently used for images of superficial layers. Final post-processing is performed by convolving an image with a σ = 0.75 pixel Gaussian kernel to reduce shot and readout noise. As the PoC prototype is still under construction all retinal images shown below have been acquired with the DCAO demonstrator.

CONE PHOTORECEPTOR IMAGING

Imaging of the cone photoreceptor layer (Fig. 10) is accomplished by focusing on deeper retinal layers. The variation in cone appearance from dark to bright in Fig. 10 is an effect of the directionality [40] or waveguide nature of the cones. The retinal photoreceptor mosaic provides all information to higher visual processing stages and is many times directly or indirectly affected or disrupted by retinal disease. It is therefore of interest to study various parameters, e.g. photoreceptor spacing, density, geometry, and size, to determine the structural integrity of the mosaic. An example of this is given inFig. 11, where the cone density of the mosaic in Fig. 10 has been calculated. Cone spacing, where possible, was obtained from power spectra of 128×128 pixel sub-regions with a 64 pixel overlap. Spacing (s) was converted to density (D) using the relation D = sqrt(3) / (2s²), and the density profile was constructed by fitting a cubic spline surface to the distribution of density values.

RETINAL CAPILLARY IMAGING

Retinal capillaries, the smallest blood vessels in the eye, are difficult to image because of their small size (down to 5 μm), low contrast, and arrangement in multiple retinal planes. Even good-quality retinal imaging fails to capture any of the finest capillary details. The preferred clinical imaging method is fluorescein angiography (FA), an invasive procedure in which a contrast agent is injected in the patient's bloodstream to enhance retinal vasculature contrast. The narrow depth of focus of both the DCAO demonstrator and the PoC prototype allows for imaging of retinal capillaries by focusing on the upper retinal layers. It is a non-invasive procedure with performance similar to FA [22]. An unfiltered camera raw image of the capillary network surrounding the fovea, the central region of the retina responsible for sharp vision, is shown in Fig. 12, and a flat-fielded image is shown in Fig. 13.

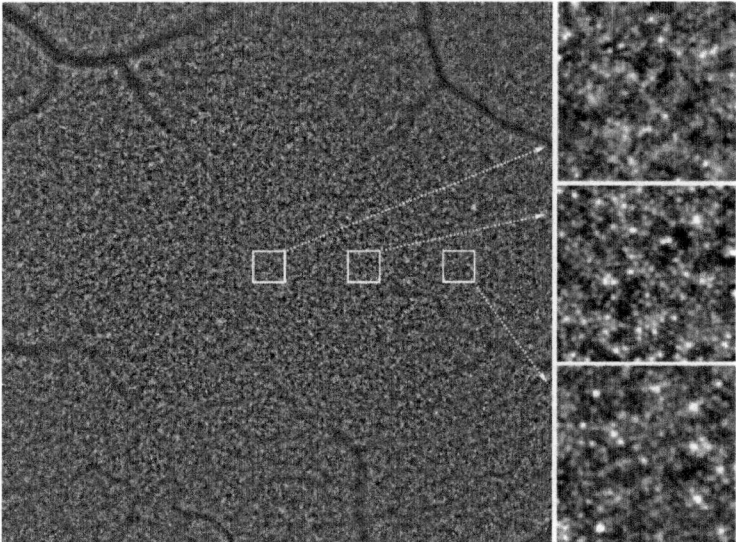

Figure 10: DCAO image of cone photoreceptor layer. Variation in cone appearance from dark to bright is an effect of the directionality or waveguide nature of cone photoreceptors.

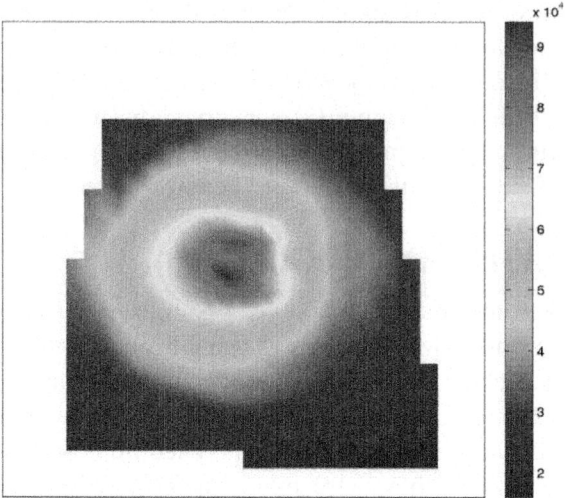

Figure 11: Cone photoreceptor density profile calculated from cone distribution in Fig. 10. Color bar represents cell density in cells/mm².

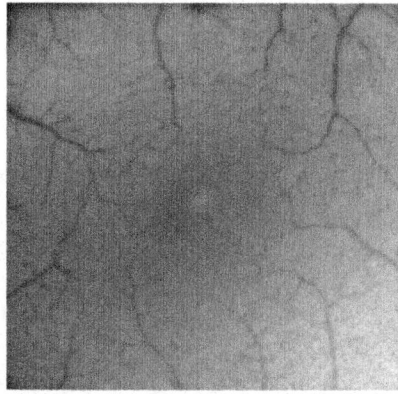

Figure 12: Camera raw DCAO image of foveal capillaries.

Nerve Fiber Layer Imaging

Evaluation of the retinal nerve fiber layer (RNFL) is of particular interest for detecting and managing glaucoma, an eye disease that results in nerve fiber loss. Changes in the RNFL are often not detectable using red-free fundus photography until there is more than 50% nerve fiber loss [41]. Although DCAO imaging does not yet provide information about RNFL thickness it can be used to obtain images with higher resolution and contrast than red-free fundus images (Fig. 14).

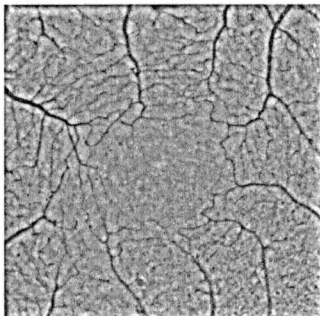

Figure 13: Image in Fig. 12 after flat-field correction. Uneven flash illumination has been reduced and retinal vessel contrast has been improved.

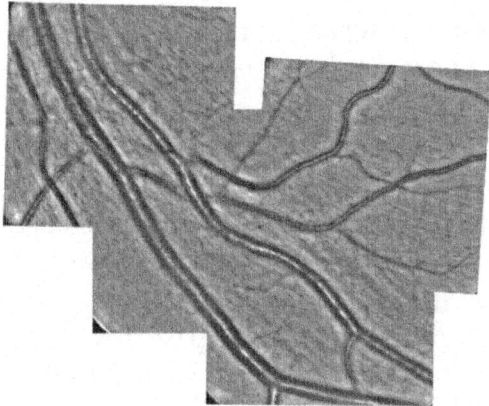

Figure 14: Montage of four DCAO images of the retinal nerve fibers and blood vessels.

CONCLUSIONS

In this chapter we have described the concept and practical implementation of dual-conjugate adaptive optics retinal imaging, i.e. multiconjugate adaptive optics using two deformable mirrors. Although the technique of adaptive optics is well established in the vision research community there are only a few publications on MCAO retinal imaging.

The DCAO instruments described here allow retinal features down to 2 μm to be resolved over a 7×7 degree FOV and enable tomographic imaging of retinal structures such as cone photoreceptors and retinal capillaries. We believe that this new technique has a future potential for clinical imaging at currently subclinical levels with an impact particularly important for early diagnosis of retinal diseases, follow-up of treatment effects, and follow-up of disease progression.

ACKNOWLEDGEMENTS

The authors would like to acknowledge financial support for this work from the Marcus and Amalia Wallenberg Memorial Fund (grant no. MAW 2009.0053) and from VINNOVA, the Swedish Governmental Agency for Innovation Systems (grant no. 2010-00518).

REFERENCES

1. Babcock. HW. The Possibility of Compensating Astronomical Seeing. Publications of the Astronomical Society of the Pacific. 1953;65(386):229.

2. Foy, Labeyrie. Feasibility of Adaptive Telescope with Laser Probe. Astronomy and Astrophysics. 1985;152(2):L29-L31.

3. Dubinin A, Cherezova T, Belyakov A, Kudryashov A. Human Retina Imaging: Widening of High Resolution Area. Journal of Modern Optics. 2008;55(4-5):671-681.

4. Dicke RH. Phase-Contrast Detection of Telescope Seeing Errors and Their Correction. Astrophysical Journal. 1975;198(3):605-615.

5. Beckers JM. Increasing the Size of the Isoplanatic Patch with Multiconjugate Adaptive Optics. ESO Conference and Workshop on Very Large Telescopes and their Instrumentation; 1988; Garching, Germany: European Southern Observatory (ESO) p. 69.

6. Beckers JM. Detailed Compensation of Atmospheric Seeing Using Multiconjugate Adaptive Optics. Roddier FJ, editor1989. 215-217 p.

7. Ellerbroek BL. First-Order Performance Evaluation of Adaptive-Optics Systems for Atmospheric-Turbulence Compensation in Extended-Field-of-View Astronomical Telescopes. Journal of the Optical Society of America a-Optics Image Science and Vision. 1994;11(2):783-805.

8. Fried DL, Belsher JF. Analysis of Fundamental Limits to Artificial-Guide-Star Adaptive-Optics-System Performance for Astronomical Imaging. Journal of the Optical Society of America a-Optics Image Science and Vision. 1994;11(1):277-287.

9. Fusco T, Conan JM, Michau V, Rousset G, Mugnier LM. Isoplanatic Angle and Optimal Guide Star Separation for Multiconjugate Adaptive Optics. In: Wizinowich PL, editor. Adaptive Optical Systems Technology, Pts 1 and 22000. p. 1044-1055.

10. Johnston DC, Welsh BM. Analysis of Multiconjugate Adaptive Optics. Journal of the Optical Society of America a-Optics Image Science and Vision. 1994;11(1):394-408.

11. Owner-Petersen M, Goncharov A. Multiconjugate Adaptive Optics for Large Telescopes: Analytical Control of the Mirror Shapes. Journal of the Optical Society of America a-Optics Image Science and Vision. 2002;19(3):537-548.

12. Rigaut FJ, Ellerbroek BL, Flicker R. Principles, Limitations and Performance of Multi-Conjugate Adaptive Optics. Adaptive Optical Systems Technology, Pts 1 and 2. 2000;4007:1022-1031.

13. Berkefeld T, Soltau D, von der Luhe O. Multi-Conjugate Adaptive Optics at the Vacuum Tower Telescope, Tenerife. Adaptive Optical System Technologies Ii, Pts 1 and 2. 2003;4839:544-553.

14. Marchetti E, Hubin N, Fedrigo E, Brynnel J, Delabre B, Donaldson R, et al. Mad the Eso Multi-Conjugate Adaptive Optics Demonstrator. Adaptive Optical System Technologies Ii, Pts 1 and 2. 2003;4839:317-328.

15. Rimmele T, Hegwer S, Marino J, Richards K, Schmidt D, Waldmann T, et al. Solar Multi-Conjugate Adaptive Optics at the Dunn Solar Telescope. 1st Ao4elt Conference - Adaptive Optics for Extremely Large Telescopes. 2009.

16. von der Luhe O, Berkefeld T, Soltau D. Multi-Conjugate Solar Adaptive Optics at the Vacuum Tower. Comptes Rendus Physique. 2005;6(10):1139-1147.

17. Rigaut F, Neichel B, Boccas M, d'Orgeville C, Arriagada G, Fesquet V, et al. Gems: First on-Sky Results. Adaptive Optics Systems III; 2012: Proc. SPIE.

18. Bedggood P, Daaboul M, Ashman R, Smith G, Metha A. Characteristics of the Human Isoplanatic Patch and Implications for Adaptive Optics Retinal Imaging. J Biomed Opt. 2008;13(2):024008. Epub 2008/05/10.

19. Bedggood P, Metha A. System Design Considerations to Improve Isoplanatism for Adaptive Optics Retinal Imaging. Journal of the Optical Society of America a-Optics Image Science and Vision. 2010;27(11):A37-A47.

20. Bedggood PA, Ashman R, Smith G, Metha AB. Multiconjugate Adaptive Optics Applied to an Anatomically Accurate Human Eye Model. Optics Express. 2006;14(18):8019-8030.

21. Thaung J, Knutsson P, Popovic Z, Owner-Petersen M. Dual-Conjugate Adaptive Optics for Wide-Field High-Resolution Retinal Imaging. Optics Express. 2009;17(6):4454-4467.

22. Popovic Z, Knutsson P, Thaung J, Owner-Petersen M, Sjostrand J. Noninvasive Imaging of Human Foveal Capillary Network Using Dual-Conjugate Adaptive Optics. Investigative Ophthalmology & Visual Science. 2011;52(5):2649-2655.

23. Chan A, Duker JS, Ko TH, Fujimoto JG, Schuman JS. Normal Macular Thickness Measurements in Healthy Eyes Using Stratus Optical Coherence Tomography. Archives of Ophthalmology. 2006;124(2):193-198.

24. Ooto S, Hangai M, Sakamoto A, Tomidokoro A, Araie M, Otani T, et al. Three-Dimensional Profile of Macular Retinal Thickness in Normal Japanese Eyes. Investigative Ophthalmology & Visual Science. 2010;51(1):465-473.

25. Barrett HH, Myers KJ. Foundations of Image Science. Hoboken, NJ: Wiley-Interscience; 2004. xli, 1540 p. p.

26. Blanc A, Fusco T, Hartung M, Mugnier LM, Rousset G. Calibration of Naos and Conica Static Aberrations - Application of the Phase Diversity Technique. Astronomy & Astrophysics. 2003;399(1):373-383.

27. Carrano CJ, Olivier SS, Brase JM, Macintosh BA, An JR. Phase Retrieval Techniques for Adaptive Optics. Adaptive Optical System Technologies, Parts 1 and 2. 1998;3353:658-667.

28. Lofdahl MG, Scharmer GB, Wei W. Calibration of a Deformable Mirror and Strehl Ratio Measurements by Use of Phase Diversity. Applied Optics. 2000;39(1):94-103.

29. Muller RA, Buffingt.A. Real-Time Correction of Atmospherically Degraded Telescope Images through Image Sharpening. J Opt Soc Am A Opt Image Sci Vis. 1974;64(9):1200-1210.

30. Murray L. Smart Optics: Wavefront Sensor-Less Adaptive Optics - Image Correction through Sharpness Maximisation. NUI Galway; 2006.

31. Ren D, Rimmele TR, Hegwer S, Murray L. A Single-Mode Fiber Interferometer for the Adaptive Optics Wave-Front Test. Publications of the Astronomical Society of the Pacific. 2003;115(805):355-361.

32. Turaga D, Holy TE. Image-Based Calibration of a Deformable Mirror in Wide-Field Microscopy. Applied Optics. 2010;49(11):2030-2040.

33. Yoon G. Wavefront Sensing and Diagnostic Uses. In: Porter J, Queener

H, Lin J, Thorn K, Awwal A, editors. Adaptive Optics for Vision Science: Principles, Practices, Design and Applications: (Wiley-Interscience; 2006. p. 63-81.

34. Rigaut F. Ground-Conjugate Wide Field Adaptive Optics for the Elts. Beyond Conventional Adaptive Optics; 2001; Venice, Italy: European Southern Observatory, Garching p. 11-16.

35. Liou HL, Brennan NA. Anatomically Accurate, Finite Model Eye for Optical Modeling. Journal of the Optical Society of America a-Optics Image Science and Vision. 1997;14(8):1684-1695.

36. Goncharov AV, Dainty JC, Esposito S. Compact Multireference Wavefront Sensor Design. Opt Lett. 2005;30(20):2721-2723.

37. Landell D. Implementation and Optimization of a Multi Conjugate Adaptive Optics Software System for Vision Research. MSc thesis. University of Gothenburg; 2005.

38. Marcos S, Moreno E, Navarro R. The Depth-of-Field of the Human Eye from Objective and Subjective Measurements. Vision Res. 1999;39(12):2039-2049. Epub 1999/05/27.

39. Howell SB. Handbook of Ccd Astronomy. Cambridge, U.K. ; New York: Cambridge University Press; 2000. xi, 164 p. p.

40. Stiles WS, Crawford BH. The Luminous Efficiency of Rays Entering the Eye Pupil at Different Points. Proceedings of the Royal Society of London. 1933;112:428-450.

41. Quigley HA, Addicks EM. Quantitative Studies of Retinal Nerve Fiber Layer Defects. Arch Ophthalmol. 1982;100(5):807-814. Epub 1982/05/01.

Chapter 9

DEVICES AND TECHNIQUES FOR SENSORLESS ADAPTIVE OPTICS

S. Bonora[1], R.J. Zawadzki[2], G. Naletto[1, 3], U. Bortolozzo[4] and S. Residori[4]

[1]CNR-IFN, Laboratory for UV and X-Ray and Optical Research, Padova, Italy

[2]VSRI, Department of Ophthalmology and Vision Science, University of California Davis, Sacramento, CA, USA

[3]Department of Information Engineering, University of Padova, Padova, Italy

[4]INLN, Université de Nice-Sophia Antipolis, CNRS, France

INTRODUCTION

Minimizing the aberrations is the basic concern of all the optical system designers. For this purpose, a large amount of work has been carried out and plenty of literature can be found on the subject. Until the last twenty years, the large majority of the optical design was related to "static" optical systems, where several opto-mechanical parameters, such as refractive index, shape, curvatures, etc. are slowly time dependent. In these systems, simple mechanisms can be adopted to change the relative position of one or more optical elements (for example, the secondary mirror of many astronomical telescopes), or slightly modify their shape and curvature (as in some synchrotron beamlines, where some optical surfaces are mechanically bent) to compensate defocusing. In the last years, a new type of optical systems, that we may call "dynamical", have heavily occupied the interest of optical designers, opening the possibility of working also in situations where the system environment varies rather quickly with time, either in a controlled or not-controlled way. For this class of optical systems adaptive optics (AO) with a closed loop control system has to be implemented. The correction of dynamical systems was predicted by Babcock in the 1953 [1] and, then, the first prototypes were realized in the early 70s with the purpose of satellite surveillance and launching high power laser beams trough the atmosphere [2].

The most known scientific applications of closed loop correction by means of AO is the acquisition of astronomical images in ground-based telescopes [3] and in-vivoimaging of cone photoreceptor mosaic by AO enhanced Fundus Cameras [4]. In astronomy, to remove the so called "seeing effect", the star light twinkling due to local dynamic variations of the atmospheric density in the air column above the telescope, it is necessary to have the real time knowledge of the wavefront of the observed object. This can be realized, for instance, by means of a Shack-Hartmann wavefront sensing device coupled to a dedicated fast algorithm which returns the mathematical description of the wavefront aberration, typically through a Zernike series decomposition [5]. Then, this information is suitably coded and passed to an AO, as a fast deformable mirror located along the optical path, that adapts its shape to compensate the time dependent aberrations. Similarly in vision science [6, 7] or retinal imaging [8-15], static and dynamic aberrations created by variation in shape of eye refractive elements and eye movements are measured by wavefront sensor, usually Schack-Hartmann and corrected by wavefront corrector, in most cases a deformable mirror. Other applications which make use of AO systems are for example: free space optical communication systems [16, 17], microscopy [18-20] or beam shaping in laser applications [21]. It is, however, rather obvious that not all AO applications have similar needs, and in particular that in some cases systems simpler than the astronomical ones can be realized. For example, in some cases there is no need to have the real time information about the aberrated wavefront: either because the aberration variation is slow [22] or because there is a specific phase which remains for a limited amount of time, as for example when correcting low order ocular aberrations "eyeglass prescription"for patient in ophthalmic diagnostics, or in optical devices in which the environmental conditions are not initially defined but the system remains stable in time [23-25]. In all these cases it can be convenient to have a simpler AO system, able to correct only the slow variations of the wavefront aberrations.

In the above mentioned cases the wavefront correction can be operated with a strong reduction in the hardware complexity, in particular by using a sensorless approach. Several techniques have been developed which use these simpler AO systems. They are generally based on the optimization of some merit function that depends on the optical system under consideration.

The algorithms for the sensorless correction can be divided into two main classes: the stochastic and the image-based ones. In the first class, the system is optimized starting from a random set and, then, applying an iterative selection of the best solutions. These algorithms have the advantage of not requiring any preliminary information about the system but they take a lot of time for converging. Many algorithms using this approach have been written and exploited successfully in different fields. Among them the most popular are: genetic algorithms [18, 26, 24], simulated annealing [13], simplex or ant colonies [27]. These approaches have the drawback of requiring a rather long computation time, or many iterations before converging, taking up to several minutes before reaching the desired system optimization.

Other sensorless techniques can be realized by analyzing some specific known feature, either intrinsic to the system or artificially introduced. An example of the latter case can be found in [28-29]. With respect to classical AO systems, the sensorless approach offers the advantage of not needing the wavefront sensor: this reduces the cost of the instrument and avoids all the problems related to maintaining the performance of such a device once installed and aligned. However, the absence of the wavefront sensor implies also some limitations, for instance, a much longer time before reaching an optimal image quality, or a final image not perfectly optimized. Clearly, the required final result and the available resources are the key elements driving the choice towards one system or another. In section 2, we will explain in detail the genetic algorithm and the ant colonies optimization process, while providing a few examples of their application in optical experimental setups.

The image-based algorithms will be explained in section 3, together with a few examples of recently reported successful applications in optical experiments. New devices useful to generate the bias aberrations will also be presented.

STOCHASTIC ALGORITHMS FOR SENSORLESS CORRECTION

GENETIC ALGORITHM

A genetic algorithm [30] searches the solution of a problem by simulating the evolution process. Starting from a population of possible solutions, it saves some of the strongest elements, that are the only ones selected to survive, and, thus, are able to reproduce themselves giving rise to the next generations. In general, the inferior individuals can survive and reproduce with a smaller probability.

This strategy allows solving a large class of problems without any initial hypothesis or preliminary knowledge. Its effectiveness was demonstrated in many experimental setups, as will be discussed in the following paragraphs.

The main steps of a genetic algorithm are depicted in Table 1 and in Fig. 1.

Table 1: Main steps required by a genetic algorithm

Starting random Population
1_selection function
2_reproduction function
3_evaluate population
4_repeat from step 1

The initial population is chosen randomly in the whole set of possible solutions. The selection function can be either probabilistic or deterministic. In the probabilistic case, the strongest elements have more chances of being selected and of reproducing to the next generation. This decreases the possibility of falling in a "local" maximum solution.

The reproduction function creates new individuals from the old population. There are two kinds of functions: crossover and mutations.

CrossOver functions: they mix the genes of the two parents by slightly modifying them and by obtaining two sons.

Example: EuristicXOver:

From the parents $V_a^{(k-1)}$ and $V_b^{(k-1)}$, the children $V_a^{(k)}$ and $V_b^{(k)}$ are generated by the following rule:

$$V_a^{(k)} = V_a^{(k-1)} + r\left(V_b^{(k-1)} - V_a^{(k-1)}\right)$$

$$V_b^{(k)} = V_b^{(k-1)}$$

Mutations functions: the genes of the parent are randomly modified.

Example: Uniform Mutation:

$$V_{cj}^{(k)} = \begin{cases} V_{cj}^{(k-1)} + w(k)\left(1 - V_{cj}^{(k-1)}\right) & \text{if } rand > 0.5 \\ V_{cj}^{(k-1)} + w(k)V_{cj}^{(k-1)} & \text{if } rand < 0.5 \end{cases}$$

Where w(k) is weight function which decreases with the iteration k.

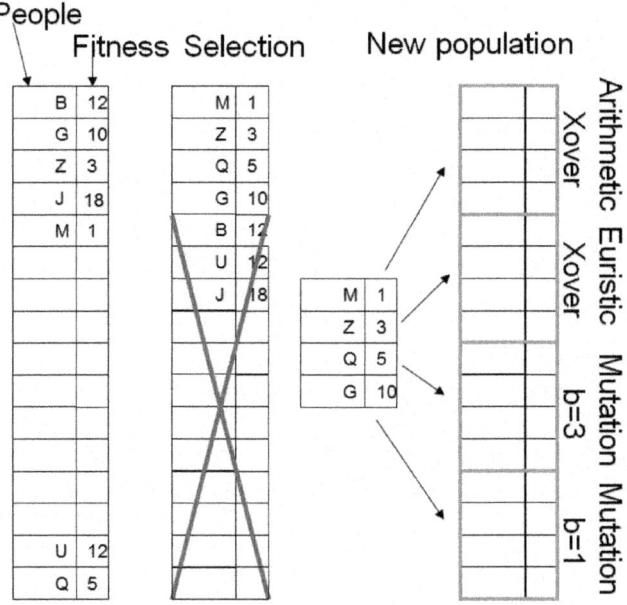

Figure 1: Diagram representing the genetic algorithm principle. The algorithm starts from a random population and then each individual is measured and the population is sorted according to its fitness. Then, some of the best individuals are selected for the generation of the next population.

Application Example: Laser Focalization

The intensity of a laser in its focal spot is largely dependent on the quality of the focal point, and this effect is even stronger in nonlinear optics. Often, in laser systems it is not simple to reach an optimal alignment, so that AO devices can be very useful in these cases.

For example, in ref. [24] it was demonstrated how an AO sensorless optimization based on a genetic algorithm can largely enhance the XUV high-order harmonics (HH) generated by the interaction of an ultrafast laser and a gas jet.

The AO system was composed by an electrostatic deformable mirror (Okotech) placed before the interaction chamber as illustrated in Fig. 2. The feedback for the genetic algorithm was the photon flux at the shortest wavelengths acquired placing a photomultiplier tube at the XUV spectrometer output.

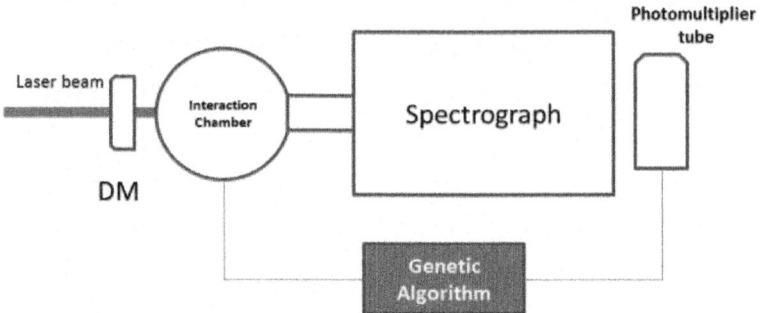

Figure 2: Experimental setup for the optimization of a laser focalization used for high order harmonics generation in ultrafast nonlinear optics. The pulsed laser beam interacts with a gas jet in the interaction chamber. The photomultiplier tube collects the signal from the spectrograph and feeds the genetic algorithm that drives the deformable mirror DM.

The laser pulse was generated by a Ti:S CPA laser system with a hollow-fiber to realize the compression of the pulse duration. The typical values used in the experiment are 6 fs of duration, 200 µJ of pulse energy, at 1 kHz repetition rate (all the experimental details are described in Villoresi et al. 2004). The focusing of the laser pulses on the gas jet, after the modifications introduced by the Deformable Mirror (DM), is obtained by means of a 250 mm focal length spherical mirror. The spectrometer that analyzes the HHs beam is based on a flat varied-line-spacing grazing-incidence grating with two toroidal mirrors.

The real-time acquisition of the spectral intensity is realized by the combination of a solar-blind open microchannel-plate (MCP) with MgF_2 photocathode and a phosphor screen placed on the spectrometer focal plane, which converts the HHs XUV spectrum in the visible, and by a photomultiplier which acquires a HHs spectral interval selected with a slit. In this way, the single-shot intensity of a single harmonic, or group of harmonics, is used as feedback by the algorithm. A separate optical channel acquires in parallel the image of all at the MCP, from which the HHs spectrum is obtained.

The genetic algorithm used a population of 80 individuals, with a deterministic selection rule that saved the 13 best ones. Both mutations and crossover were used. The results showed an increase of the XUV photons by a factor of 5 when the algorithm was applied. Moreover, the cutoff region moved to shorter wavelengths as reported in Fig 3. The optimization process took about 20 iterations to converge.

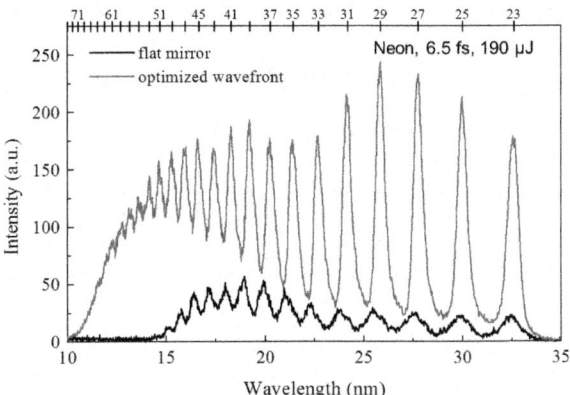

Figure 3: Result of the experimental optimization of the high order harmonics genera-tion spectra in the case of the flat AO mirror (black line) and in the case of the opti-mized wavefront (red line).

Ant Colonies

Ant colonies, in natural world, search the food by walking randomly. After having found it, they return to their colony leaving down a pheromone trail. If other ants cross the same trail they will not walk randomly but they will likely follow it and will reinforce the pheromone trail. The more ants will find food at the end of the trail, the more pheromone will mark it. However, since the pheromone evaporates reducing its strength, the described process will make the shortest path which will be the one with the highest density of pheromone, so providing a selection among all the possible paths, as illustrated in Fig. 4.

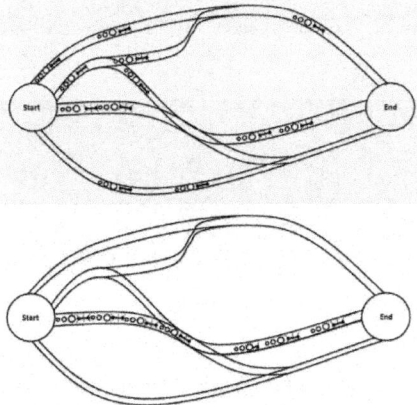

Figure 4: Ants start randomly their search for food, then the shortest path gets the higher content of pheromone. Finally, the ants will follow with larger probability the path having the highest content of pheromone.

The main essence of the Ant Colonies optimization algorithm [27] is to simulate the ant behavior for the optimization of a given problem. The algorithm steps necessary for running the optimization are listed in Table 2.

Table 2: Steps of an ant colony algorithm

1_Set the initial ants position on the trail
2_Compute the paths length
3_Update pheromone
4_Move the ants
5_Go to step 2

As an example we show in Fig. 5 the simulation of the application of the ant colony strategy to a deformable mirror with 32 actuators and 8 bits control. In this example the actuators and their control values are the domain in which the ants can move. In the simulation the shortest path is a parabolic function, which is represented by the red line. Fig. 5 (top) shows the initial random pheromone distribution, while Fig. 5 (bottom) shows the pheromone distribution at the end of the optimization process.

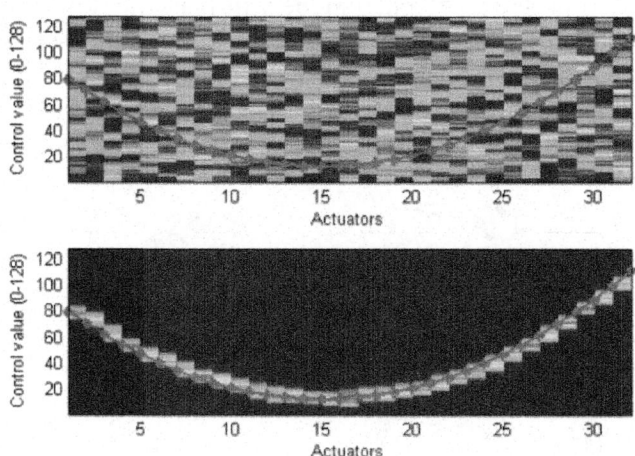

Figure 5: Implementation of an ant colony strategy for the optimization of a deformable mirror with 32 actuators and 8 bit control. The red curve represents the shortest (optimized) path. The top panel shows the initial random pheromone distribution while the bottom panel shows the pheromone at the end of the selection process.

Application Example: Quantum Optics

The quality of an optical wavefront plays an important role in Spontaneous Down Conversion (SPDC) process. As demonstrated by [31] the use of a deformable mirror can enhance the generation of photon pairs acting on the wavefront before the generation takes place in the nonlinear crystal. In that system the optimization was carried out by the use of an electrostatic DM (PAN, Adaptica srl) and the application of the ant colonies algorithm.

In the experiment, the pump beam is reflected by the DM to a BBO type-I nonlinear crystal. Then, the degenerate SPDC photons at 808 nm are selected and measured by a high efficiency SPADs (Single Photon Avalanche Diode). Since the wavefront has a strong effect on the downconverted light, it can strongly affect the coupling in the fibers of the SPAD detectors. The feedback for the algorithm imposed the condition of photon coincidences. It was demonstrated in the experiment that the coincidences rate was increased by about 20% when the optimization algorithm was applied. The algorithm used about 80 ants and the convergence took place in about 800 iterations.

IMAGE BASED ALGORITHMS

Although the stochastic optimization algorithms have been demonstrated to represent important tools for optical experiments, new techniques, which demonstrated to be more effective, have recently been introduced. The use of a modal approach, based on the application of bias aberrations and of a suitable metrics, sorted out some of the limitations of the search algorithms, such as the long convergence time and the need of a training for the determination of the algorithm parameters. This new approach demonstrated to be effective both in visual optics and in laser optimization, as described later in this section. The arbitrary generation of aberrations can be achieved through the use of deformable mirrors, either thanks to a preliminary calibration of them or through the design of a suitable new class of wavefront correctors [32].

DEVICES FOR SENSORLESS MODAL CORRECTION

Electrostatic membrane deformable mirrors rely on the electrostatic pressure between an actuator pad array and a thin metalized membrane [33]. Thus, the more the actuators the better the wavefront resolution that the mirror can control. The use of these deformable mirrors is, then, subjected to the acquisition of the deformation generated by each electrode. On the other hand, this kind of DMs can also be used with the optimization algorithms. The drawback, in this case, is that the higher the number of actuators the longer will take to the algorithm to converge.

Recently, a new type of deformable mirrors suitable for the direct generation of aberrated wavefronts was designed. The modal membrane deformable mirror, MDM, relies on the use of a graphite layer electrode arrangement (see Fig. 6) for the generation of a continuous distribution of the electric field which allows the generation of the low order aberrations (defocus, astigmatism, coma) and of the spherical aberration.

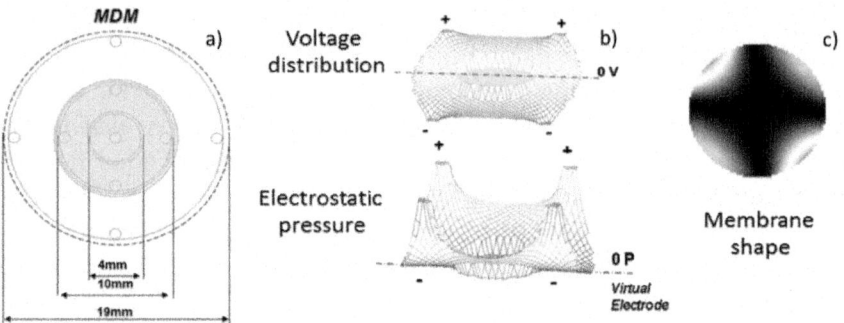

Figure 6: Electrostatic modal membrane deformable mirror, MDM. (a) Layout of the electrodes of the MDM; (b) voltage and electrostatic pressure distribution which generates the astigmatism shape illustrated in the interferogram shown in (c).

The MDM has already been demonstrated to be effective in several fields, as laser focalization [32], image sharpening and Optical Coherence Tomography (OCT), as it will be discussed later.

Another device for the generation of aberrations is the PhotoControlled Deformable Mirror (PCDM), which is schematically represented in Fig. 7. This deformable mirror [35, 36] is composed of an electrostatic membrane while the actuator pad array is replaced by a photoconductive material. Thus, the membrane shape depends on the light pattern projected on the photoconductor. Arbitrary actuator pads can be conveniently achieved by illuminating the photoconductive side of the mirror with a commercially available Digital Light Processing (DLP) hand-held projector.

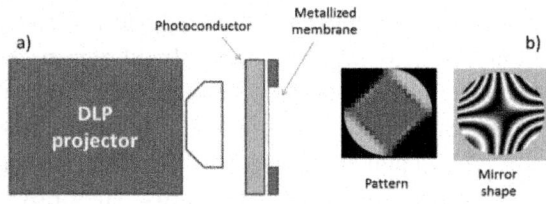

Figure 7: Photo-controlled deformable membrane mirror, PCDM. a) Schematic representation of the PCDM and the projection system allowing to achieve arbitrary ac-

tuator pads. b) Left: layout of the electrode pattern; right: correspondingly generated mirror shape; as an example the electrode pattern was chosen to generate astigmatism.

The calculation of the electrode pattern that generates a determined aberration is composed of the following steps:

division of the projector area into small subsets (i.e. 40 × 40) calculation of the membrane shape for each of the 40 × 40 pixels, solving the Poisson equation by the iterative methods;

determination of the pattern by pseudoinversion of the matrix determined at point b.

A few examples of the realized electrode patterns are shown in Fig. 8, together with the corresponding measurements of the aberrated wavefronts.

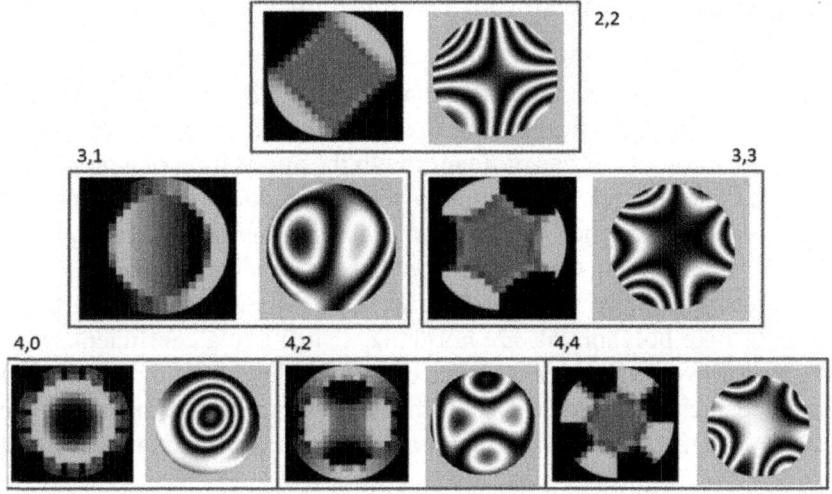

Figure 8: Generation of the first four Zernike orders with the photocontrolled deformable mirror; the light patterns necessary for their generation are on the left; the obtained corresponding interferograms are on the right.

We proved that the image quality can be considerably improved by using these adaptive devices in an image sharpening setup. For example, the MDM allowed achieving a significant image sharpening with just about 35 measurements, as illustrated in Fig. 9.

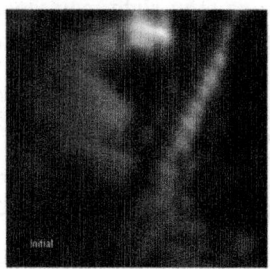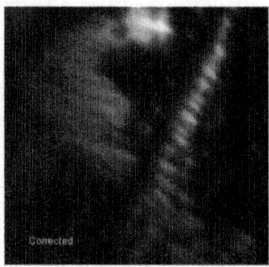

Figure 9: Optimization of an image deteriorated by aberrations; left: initial image; right: image corrected by the MDM (after 35 measurements).

Optimization Of Low Spatial Frequencies

The sharpness of an optical imaging system depends strongly on the wavefront quality. Recently [28] demonstrated that the low spatial frequency in an image can be used as a metric to perform the optimization. The process takes place by the acquisition of a series of images with the application of a predetermined aberration. The images, then, contain the information about the corrections which have to be applied to cancel the aberrations. This technique is very powerful, especially if coupled to the Lukosz modes aberration expansion. The Lukosz modes are similar to the Zernike polynomials: the difference is that the Zernike polynomials are normalized such that a coefficient of value 1 generates a wavefront with a variance of 1 rad^2, while the Lukosz functions are normalized such that a value 1 coefficient corresponds to a rms spot radius of $\lambda/(2\pi NA)$, where λ is the wavelength and NA is the numerical aperture of the focusing lens.

The peculiarity of this expansion is that the effect of the Lukosz polynomials coefficients $\{ai\}$, on the image sharpness $I(ai)$ is quadratic:

$$I(a_i) \approx \sum_i a_i^2.$$

This implies that the optimization of each mode can be performed independently and requires just the acquisition of three images. Then, the best point for each aberration can be found by interpolating the result with a quadratic function.

Point Spread Function (Psf) Optimization

Another example of application of wavefront sensorless AO [29] consists in projecting a known point-like source through the optical system under test and then analyzing its image by means of a suitable software [36]. With the information obtained by the analysis of the point source image, the shape of a deformable mirror inserted along the optical path is modified. This process is iteratively repeated through a defined hierarchy, to gradually remove the optical aberrations.

With this technique, the point source image to be analyzed is not directly available and has to be somehow created. As an example, in the case of a fundus camera dedicated to the observation of the human retina, an illuminated pinhole can be projected on the retina itself through a dedicated optical path; this is a standard technique for this type of applications and is not going to introduce a significant complexity in the system. The light that is back-diffused by the retinal fundus acts as a point source, and its wavefront can be analyzed to estimate the aberrations present along the optical path from the retina to the detector.

Application Of The Psf Optimization In A Visual Optics Setup

The closed loop method of correcting the aberrations of an optical system has been verified to be very stable, at least with respect to possible misalignments of the deformable mirror or aging of the mirror membrane that has been used. This stability is inherent in the adopted approach to the problem, which is less ambitious than correcting the wavefront aberration.

The described technique has been verified by means of the rather simple optical setup shown in Fig. 10. The radiation emitted by a LED diode source (SOU) is condensed by a microscope objective lens (L_{cond}) on a pinhole (PH). The radiation emerging from the pinhole is collected by a zoom collimating lens (L_{coll}). The collimated beam passes through a diaphragm (DIA) and a beam splitter (BS) and impinges normally onto a deformable mirror (M_{def}). After reflections from M_{def} and BS, the beam is compressed by an a-focal Newtonian system (L_{comp}^1 and L_{comp}^2) and can, then, follow two different paths: either a) a focusing two-lens system L_{foc} that makes the image of the pinhole on a CMOS digital camera (DET), or b) a flip mirror (M_{flip}) which deviates the beam on a wavefront analyzer (WFA). The latter has been used to measure the wavefront aberrations before and after the correction performed by the DM. With this system, both by varying the focal length of L_{coll} and tilting L_{comp}^1, it was possible to introduce controlled amounts of aberrations on the nominal pinhole image. Then, by the suitable image analysis and consequent estimate of the aberrations, the parameters needed to drive the deformable mirror to improve the image quality have been derived.

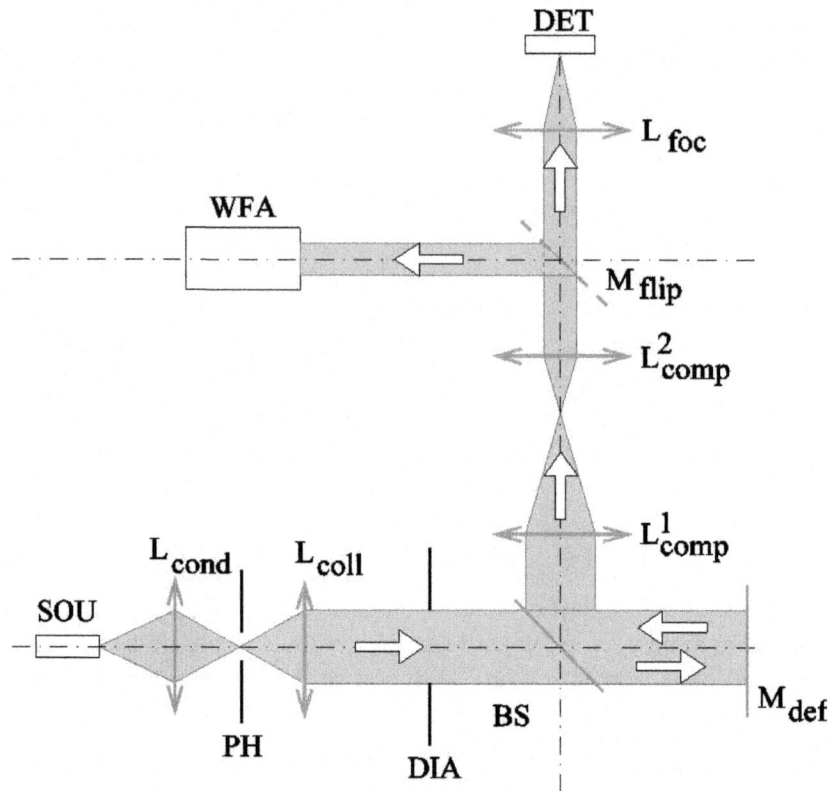

Figure 10: Schematic representation of the optical setup used for testing the capability of correcting the system aberrations with a sensorless technique. SOU: source LED diode; M_{def}: deformable mirror; DET: CMOS camera for detection; WFA: wavefront analyzer. See text for a complete description.

Even if the apparatus performance was constrained by the limited unidirectional sag of the deformable mirror, the obtained results proof the principle of the adopted methodology. This is clearly demonstrated in Fig. 11, which shows the wavefront error measured with the WFA before and after the deformable mirror correction for three different cases. From these graphs, and more quantitatively from the detailed analysis described in [29], it can be seen that a RMS wavefront error as low as /10 (@527.5 nm) can be obtained, which is a significant result for a sensorless AO system. The correction was not particularly effective only in those cases in which the unidirectionality of the mirror deformation did not allow aberration compensation, as in case of astigmatism. However, with a different choice of AO system, the system is very effective in identifying and correcting aberrations.

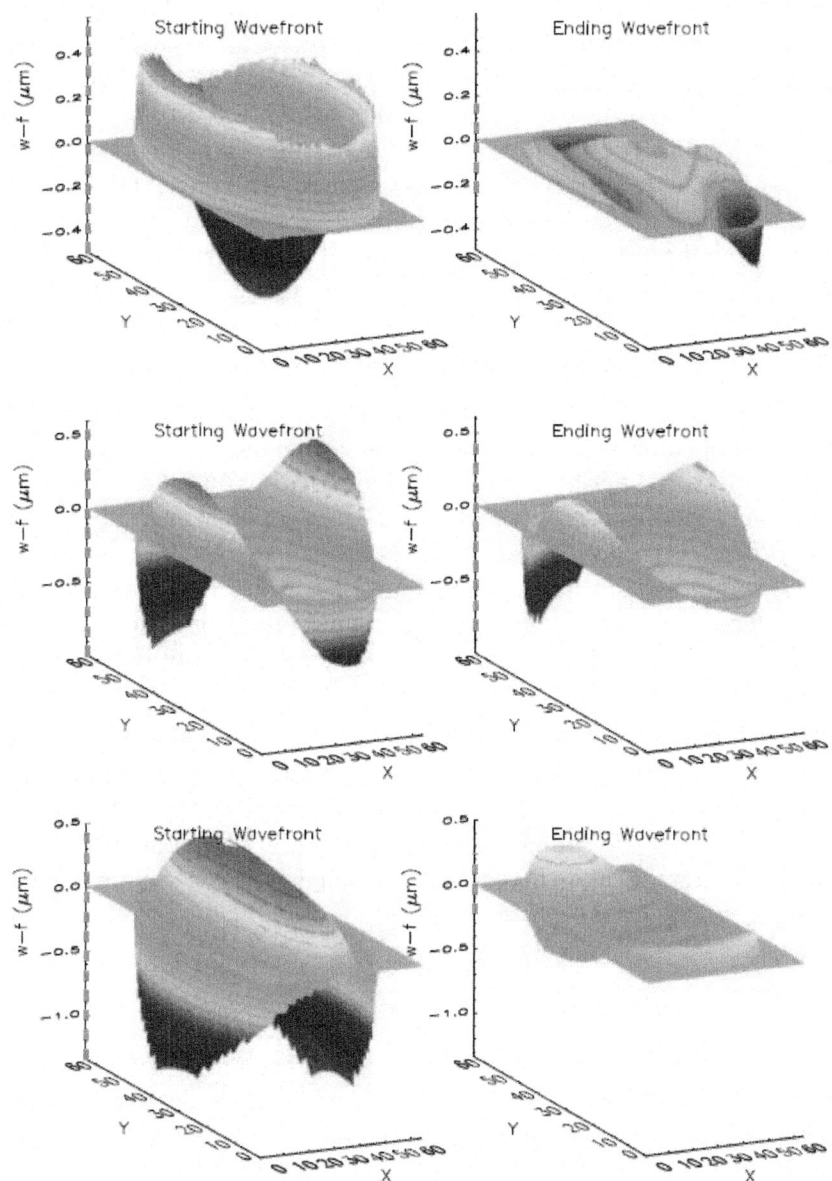

Figure 11: Wavefront plots (obtained with the wavefront analysis, WFA) before and after the correction applied by the deformable mirror for three considered cases. Top: the main aberration is defocus; Middle: the main aberrations are astigmatism and coma; Bottom: the main aberrations are defocus and astigmatism. The blue dashed lines over plotted to the Z axis represent the total wavefront excursion.

It has to be mentioned that these tests verified that the image analysis algorithm takes less than 100 cycles to reach the optimal condition; since one cycles takes approximately 1/20 - 1/25 s on a standard computer, the whole optimization takes just 4-5 s. Comparing this time with the typical times necessary to optimize other sensorless AO systems, it is evident the significant advantage of this technique, once implemented. The only limitation of this technique is that starting PSF image should give enough signal. In fact, if the point source image is too spread out, the signal to noise ratio can be very poor, substantially inhibiting the system to make a correct image analysis.

Optical Coherence Tomography (Oct)

Optical coherence tomography, OCT, is an imaging modality allowing acquisition of micrometer-resolution three-dimensional images from the inside of optical scattering media (e.g. biological tissue). OCT is analogous to ultrasound imaging, except that it makes use of light instead of sound. It relies on detecting interferometric signal created by the light back scattered from the sample and from a reference arm in a Michelson or Mach-Zehnder interferometer. OCT has many applications in biology and medicine and can be treated as a sort of optical biopsy without requirement of tissue processing for microscopic examination.

One of the interesting features of OCT is that, unlike in most optical imaging techniques, the axial and lateral resolutions are decoupled, thus allowing for an improved axial resolution, which is independent of transverse resolution. The axial resolution Δz is determined by the roundtrip coherence length of the light source and can be calculated from the central wavelength (λ_0) and the bandwidth $(\Delta\lambda)$ of the light source as [37]:

$$\Delta z = \frac{2\ln 2}{\pi}\frac{\lambda_0^2}{\Delta\lambda}.$$

The lateral resolution (Δx) in OCT is defined similarly to the confocal scanning laser ophthalmoscopy (cSLO), since OCT is based on a confocal imaging scheme. In many imaging systems, however, the confocal aperture exceeds the size of the Airy disc, which degrades the resolution to the value known from microscopy, i.e. [38]:

$$\Delta x = 1.22\lambda\frac{f}{D}.$$

Therefore, as for standard microscopy, AO enhanced devices might be necessary to achieve diffraction limited transverse resolution. As a result, only a combination of OCT with AO has the potential to achieve high and isotropic volumetric resolution. The use of broadband light sources that are necessary for OCT and the complexity of both the AO and the OCT technique, make the combination very challenging [39]. In general, any AO-OCT instrument can be divided into two subsystems: an adaptive optics subsystem, with wavefront sensing and wavefront correction, and an interferometric OCT subsystem. In every implementation of AO-OCT all the elements of the AO subsystem are located in the sample arm of the OCT interferometer. Indeed, there is no need to have AO correction in the reference arm because aberrations introduced within this part of the system will not influence the transverse resolution of the image. In most of the AO-OCT systems, a Shack–Hartmann wavefront sensor is used to measure aberrations and, then, to control adaptive optics correction.

Bonora and Zawadzki recently demonstrated that sensorless correction can be implemented in optical coherence tomography by using a specially developed resistive deformable mirror. This novel modal deformable mirror, MDM, was successfully employed in the UC Davis AO-OCT system to image static samples, test targets and tissue phantoms. Fig. 12 shows a schematic representation of the sensorless AO-OCT system used in the experiments.

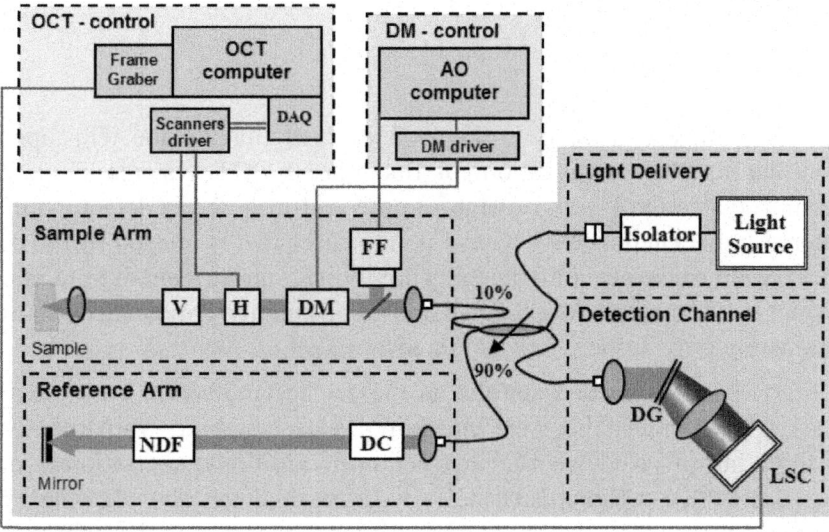

Figure 12: Schematic representation of the system for sensorless adaptive optics - optical coherence tomography. Note that there is no wavefront sensor in the sample arm. The far-field camera (FF) is used to check if the AO correction generates improved focal spots. DM : deformable mirror; V: vertical mirror galvanometer; H: horizontal mirror galvanometer. In the reference arm: NDF is a neutral density filter. The detection channel comprises a grating (DG) and a linear CCD detector (LSC). The quality of the image acquired with the OCT detection channel is used to search for DM shapes

that correct aberrations in the imaged sample. The imaging system used to acquire the data was developed in the Vision Science and Advanced Retinal Imaging Laboratory (VSRI). Details of the OCT system components can be found in [40]. Here, we briefly describe the main characteristics of the system. In the current configuration, the light source for OCT was a superluminescent diode (Broadlighter) operating at 836 nm and with a 112 nm spectral bandwidth (Superlum LTD), allowing to achieve a 3.5 μm axial resolution. The beam diameter at the last imaging objective was 6.7 mm, allowing for up to 10 μm lateral resolution when a 50 mm focal length imaging objective was used. The AO correction was optimized by using the intensity of the AO-OCT en-face projection views during the volumetric data acquisition. In the current system configuration, we have used about 9 mm diameter of the modal deformable mirror. The light reflected from the sample is combined with the light from the reference mirror, and then sent to a spectrometer. There, a CCD line detector acquires the OCT spectrum.

To test the performance of our sensorless AO-OCT system, we evaluated the image quality of a sample, consisting of a USAF resolution test chart with an adhesive tape glued to its front side, after insertion of a trial lens with 0.5 Diopter astigmatism in front of the imaging objective. We were able to achieve improved resolution by using the following merit function S [41] on the OCT en-face projection images

$$S = \int I^2(x, y)dxdy,$$

where I(x,y) is the intensity in the OCT en-face image plane. This approach is simillar to PSF optimization. In fiber based OCT systems single mode fiber introduces OCT beam to the sample and also act as detector for back scattered light. Therefore we have a point source that is imaged by the optical system and the confocal pinhole that allows direct mesurment of light intensity trougput by the system. As expected, the algorithm performed the optimization by adjusting only defocus and astigmatism (see Fig. 13).

Fig. 14 shows some examples of the en-face projection views extracted from OCT volumes: there are the initial view acquired from the sample, and three improved views after correction of additional aberrations, namely, defocus and two astigmatisms. Clearly, at each correction step the images of the test target get sharper. Additionally, the features of the adhesive tape attached to the back of the Air Force test target become more visible as well.

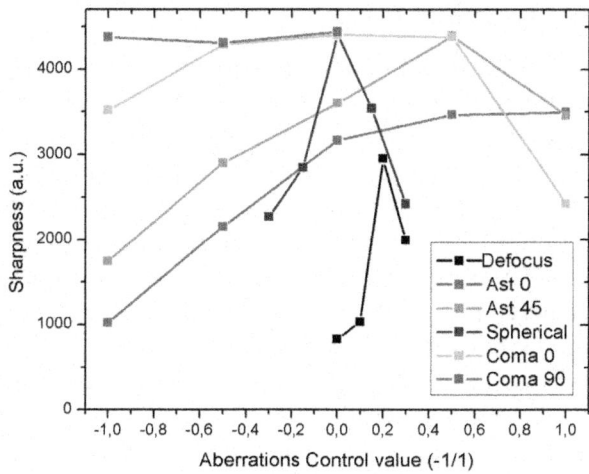

Figure 13: Graph of the Merit Function of AO-OCT images for different values of aberrations generated by the modal deformable mirror. Note that higher values correspond to better AO-corrections.

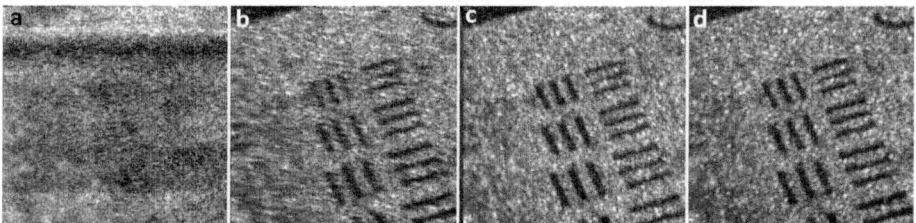

Figure 14: En-face projection views of the AO-OCT images of the test target for the best corrected values of the Zernike coefficients; (a) before correction, (b) after defocus correction, (c) after defocus and Ast 0° correction, (d) after defocus, Ast 0° and Ast 45° correction.

These recent results demonstrate that wavefront sensorless control is a viable option for imaging biological structures for which AO cannot establish a reliable wavefront that could be corrected by a wavefront corrector. Future refinements of this technique, beyond the simple implementation presented in this chapter, should allow its extension to in-vivo applications. An example of sensorless adaptive optics scanning laser ophthalmoscopy (AO-SLO) for imaging in-vivo human retina has been recently presented [42].

Laser Process Optimization

Similarly to the optimization process presented in section 2.1.1 [24], we report here about the optimization of a laser process by the use of a sensorless AO [43]. In the former case, the generation of harmonics from an ultrafast laser was improved by the use of a genetic algorithm. In the latter case, an algorithm derived from the image-based procedure was employed in conjunction with the use of a MDM deformable mirror similar to the one described in section 3.1. The advantages in terms of experimental complexity and convergence time are discussed in the given reference. In the sensorless case, the laser source was a tunable high energy mid-IR (1.2μm-1.6μm) optical parametric amplifier with 10 Hz repetition rate [44]. The harmonics of the laser were generated by the interaction of the laser pulses with a krypton gas jet. In this system, the infrared pulses and the slow repetition rate made inconvenient, respectively, the use of a wavefront sensor and of an optimization algorithm needing hundreds of iterations. The experimental setup used for this application is illustrated in Fig. 15. To demonstrate the easiness of integrating the sensorless AO device within the experiment, the optical path before the DM is shown with a dotted line. The additional elements are simply a plane mirror and a resistive MDM, which have been introduced without any complex operations. The system optimization consisted in the increase of the harmonic signal detected by the photomultiplier at the output of the monochromator. The obtained result is illustrated in Fig. 16, where it is possible to see that the photon flux on the photomultiplier is doubled with respect to the one obtained after the correction of the defocus.

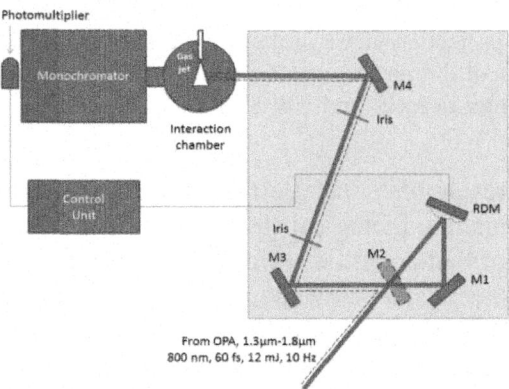

Figure 15: Experimental setup for the generation of harmonics from a femtosecond tunable high-energy mid-IR optical parametric amplifier, OPA. Dotted line: optical path before the insertion of the MDM. Red line: optical path realized for the experiment with the deformable mirror.

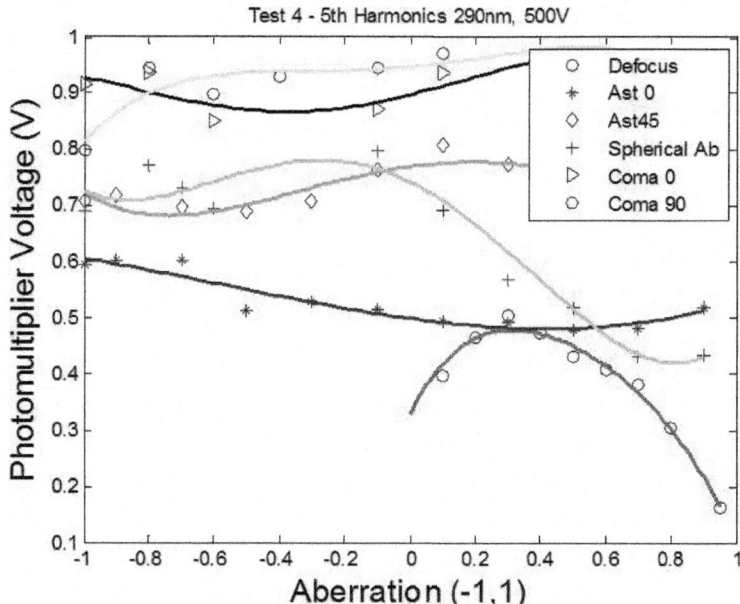

Figure 16: Optimization of the voltage generated by the photomultiplier over a 50 Ω load for the 5th harmonic at 290 nm, obtained by the use of krypton gas.

CONCLUSIONS

In adaptive optics the choice of the optimal correction strategy depends on the required application, desired image quality, and affordable complexity/cost of the final system. In this context, sensorless adaptive optics provides several solutions, most of them implementable at a simplified and relatively low-cost level, that can be exploited for a wide range of applications.

We have presented here both a review of the most diffused systems used in sensorless adaptive optics and some recently developed algorithms and devices. Essentially, two different approaches are employed: those based on random search and the subsequent application of evolutionary strategies, and those based on the application of some bias aberration. In general, the second class of algorithms present a faster convergence.

We have shown several application examples in different fields, such as the optimization of ultrafast nonlinear optical systems for the generation of high order harmonics, the image sharpening in microscopy applications and the enhancement of optical coherent tomography.

Sensorless adaptive optics appears, therefore, as having a great potential for finding new applications in current and future technologies. The continuous improvement of the optimization algorithms and development of novel deformable mirror devices, make the integration of AO into various optical systems increasingly easier. Particularly, the conjunction of sensorless AO with OCT might open the way to a new generation of diagnostic imaging.

REFERENCES

1. Babcock H.W., The Possibility of Compensating Astronomical Seeing, Publication of the Astronomical Society of the Pacific Vol.65, No. 386, pp. 229, (1953).

2. Tyson R., Tharp J., Canning D., Measurement of the bit-error rate of an adaptive optics, free-space laser communications system, part 2: multichannel configuration, aberration characterization, and closed-loop results, Optical Engineering Vol. 44, No. 9, pp. 096,003-1 096,003-6, (2005).

3. Hardy J.W., Adaptive Optics for Astronomical Telescopes, (Oxford University Press, ISBN-10: 0195090195, USA, 1998).

4. Liang J., Williams D., Miller D., Supernormal vision and high resolution retinal imaging through adaptive optics, Journal of the Optical Society of America A, Vol. 14, No. 11, pp. 2884-2892, (1997).

5. Irwan R., Lane R., Analysis of optimal centroid estimation applied to Shack Hartmann sensing, Applied Optics Vol. 38, No. 32, pp. 6737-6743, (1999).

6. Porter J. , Queener H. , Lin J., Thorn K. E., Awwal A., Adaptive Optics for Vision Science: Principles, Practices, Design and Applications (Wiley, 2006).

7. Roorda A., Adaptive optics for studying visual function: A comprehensive review, Journal of Vision, Vol. 11, No. 5, pp. 1-21 (2011).

8. Liang J., Williams D., Aberrations and retinal image quality of the normal human eye, Journal of the Optical Society of America A, Vol. 14, No. 11, pp. 2873-2883, (1997).

9. Zhu L., Sun P., Bartsch D., Freeman W., Fainman Y., Adaptive control of a micro-machined continuous-membrane deformable mirror for aberration compensation, Applied Optics, Vol. 38, No. 1, pp. 168-176, (1999).

10. Le Gargasson J.-F., Glanc M., Léna P., Retinal imaging with adaptive optics, ComptesRendus de l›Acadèmie des Sciences - Series IV - Physics 2, pp. 1131-1138, (2001).

11. Roorda A., Romero-Borja F., Donnelly W., Queener H., Hebert T., Campbell M., Adaptive optics scanning laser ophthalmoscopy, Optics Express, Vol. 10, No. 9, pp. 405-412, (2002).

12. Zawadzki R., Jones S., Olivier S., Zhao M., Bower B., Izatt J., Choi S., Laut S., Werner J, Adaptive-optics optical coherence tomography for high-resolution and high-speed 3D retinal in vivo imaging, Optics Express, Vol. 13, No. 21, pp. 8532-8546, (2005).

13. Zommer S., Ribak E., Lipson S., Adler J., Simulated annealing in ocular adaptive optics, Optics Letters, Vol. 31, No. 7, pp. 1-3, (2006).

14. Gray D., Merigan W., Wolfing J., Gee B., Porter J., Dubra A., Twietmeyer T., Ahamd K., Tumbar R., Reinholz F., Williams D., In vivo fluorescence imaging of primate retinal ganglion cells and retinal pigment epithelial cells, Optics Express, Vol. 14, No. 16, pp. 7144-7158, (2006).

15. Fernandez E., Vabre L., Adaptive optics with a magnetic deformable mirror: application in the human eye, Optics Express, Vol. 14, .No. 20, pp. 8900-8917, (2006).

16. Tyson R., 1999, Adaptive Optics Engineering Handbook , CRC Press, ISBN-10: 0824782755, New York USA

17. Tyson R., Tharp J., Canning D., Measurement of the bit-error rate of an adaptive optics, free-space laser communications system, part 1: tip-tilt configuration, diagnostics, and closed-loop results, Optical Engineering Vol. 44, No. 9, pp. 096,002-1 096,002-6, (2005).

18. Albert O., Sherman L., Mourou G., and Norris T., Smart microscope: an adaptive optics learning system for aberration correction in multiphoton confocal microscopy, Optics Letters Vol. 25, No. 1, pp. 52-54, (2000).

19. Neil MAA., Juskaitis R., Booth M.J., Wilson T., Tanaka T., Kawata S., Adaptive Aberation correction in two-photon microscope, Journal of microscopy, Vol 200, pp 1-5-108 (2000).

20. Booth M.J., Neil M.A.A., Juskaitis R., Wilson T., Adaptive aberration correction in a confocal microscope, PNAS, Vol. 99, No. 9. Pp. 5788-5792 (2002).

21. Brida D., Manzoni C., Cirmi G., Marangoni M.,, Bonora S., Villoresi P., De Silvestri S., Cerullo G., (2010), Few-optical-cycle pulses tunable from the visible to the mid-infrared by optical parametric amplifiers, Journal of Optics, Vol. 12, No. 1, (January 2010), 2040-8978

22. Okada T., Ebata K., Shiozaki M., Kyotani T., Tsuboi A., Sawada M., Fukushima M., Development of adaptive mirror for CO2 laser, in High-Power Lasers in Manufacturing, X. Chen, T. Fujioka, and A. Matsunawa,

eds., Vol. 3888 of SPIE Proc., pp. 509-520, (2000).

23. Jackel S., Moshe I., Adaptive compensation of lower order thermal aberrations in concave-convex power oscillators under variable pump conditions, Optical Engineering Vol. 39, No. 09, pp. 2330-2337, (2000).

24. Villoresi P., Bonora S., Pascolini M., Poletto L., Tondello G., Vozzi C., Nisoli M., Sansone G., Stagira S., De Silvestri S., Optimization of high-order-harmonic generation by adaptive control of sub-10 fs pulse wavefront, Optics Letters, Vol. 29, No.2, pp. 0146-9592, (2004).

25. Zacharias R., Beer N., Bliss E., Burkhart S., Cohen S., Sutton S., Atta R.V., Winters S., Salmon J.T., Stolz M. L. C., Pigg D., Arnold T., Alignment and wavefront control systems of the National Ignition Facility, Optical Engineering, Vol. 43, No. 12, pp. 2873-2884, (2004).

26. Gonté F., Courteville A., Dandliker R., Optimization of single-mode fiber coupling efficiency with an adaptive membrane mirror, Optical Engineering, Vol. 41, No. 5, pp. 1073-1076, (2002).

27. Bonabeau E., Dorigo M. and Theraulaz G., Inspiration for optimization from social insect behaviour, Nature Vol. 46, pp. 39-42, (2000).

28. Debarre D., Booth M.J. and Wilson T., Image based adaptive optics through optimisation of low spatial frequencies, Optics Express, Vol. 15, No. 13, pp. 8176-8190, (2007).

29. Naletto G., Frassetto F., Codogno N., Grisan E., Bonora S., Da Deppo V., Ruggeri A. (2007), No wavefront sensor adaptive optics system for compensation of primary aberrations by software analysis of a point source image, Part II: tests, Applied Optics Vol. 46, No. 25, pp. 6427-643, 0003-6935, (2007).

30. Judson R.S., Rabitz H., Teching lasers to control molecules, Phys. Rev. Lett., Vol. 68, No. 10, pp. 1079-7114 (1992).

31. Minozzi M., Bonora S., Vallone G., Segienko A., Villoresi P., Bi-photon generation with optimized wavefront by means of Adaptive Optics, 11th International Conference on quantum communication, Vienna, Austria, 30 July-3 August, 2012.

32. Bonora S., Distributed actuators deformable mirror for adaptive optics, Optics Communications, Vol. 284, No. 13, pp. 0030-4018 , (2011).

33. Bonora S., Capraro I., Poletto L., Romanin M., Trestino C., Villoresi P., Fast wavefront active control by a simple DSP-Driven deformable mirror, Review of Scientific Instruments, Vol. 77, No. 9, pp. 0034-6748 ,(2006).

34. Bortolozzo U., Bonora S., Huignard J.P., Residori S., Continuous photocontrolled deformable membrane mirror, Applied Physics Letters,

Vol. 96, No.25, pp. 0003-6951, (2010).

35. Bonora S., Coburn D., Bortolozzo U., Dainty C., Residori S., High resolution wavefront correction with photocontrolled deformable mirror, Optics Express Vol. 20, No. 5, pp. 5178-5188, (2012).

36. Grisan E., Frassetto F., Da Deppo V., Naletto G., Ruggeri A., No wavefront sensor adaptive optics system for compensation of primary aberrations by software analysis of a point source image. Part I: methods, Applied Optics, Vol. 46, No. 25, pp. 6434-6441, 0003-6935, (2007).

37. Fercher AF, Hitzenberger CK. Optical coherence tomography. In: Progress in Optics, Vol. 44, Chapter 4, pp. 215-302, Wolf E. Editor, (Elsevier Science & Technology, 2002).

38. Zhang Y, Roorda A. Evaluating the lateral resolution of the adaptive optics scanning laser ophthalmoscope, J. Biomed. Opt., Vol. 11, No. 1, pp. 014002, (2006).

39. Pircher M., Zawadzki R.J. , Combining adaptive optics with optical coherence tomography: Unveiling the cellular structure of the human retina in vivo, Expert Review of Ophthalmology, Vol. 2, No. 6, pp. 1019-1035, (2007).

40. Zawadzki R.J., Jones S.M., Pilli S., Balderas-Mata S., Kim D., Olivier S.S., Werner J.S., Integrated adaptive optics optical coherence tomography and adaptive optics scanning laser ophthalmoscope system for simultaneous cellular resolution in vivo retinal imaging, Biomed. Opt. Express Vol. 2, No. 6, pp. 1674-1686, (2011).

41. Muller R. A., Buffington A., Real-time correction of atmospherically degraded telescope images through image sharpening, J. Opt. Soc. Am., Vol. 64, No. 9, pp. 1200–1210, (1974).

42. Hofer H., Sredar N., Queener H., Li C., Porter J., Wavefront sensorless adaptive optics ophthalmoscopy in the human eye, Opt. Express Vol. 19, No. 14160-14171, pp.14160-14171, (2011)

43. Bonora S., Frassetto F., Coraggia S., Spezzani C., Coreno M., Negro M., Devetta M, Vozzi, C., Stagira, S. Poletto L., Optimization of low-order harmonic generation by exploitation of a deformable mirror, Applied Physics B, Vol. 106, No. 4, pp.905-909, (2011).

44. Vozzi C., Calegari F., Benedetti E., Gasilov S., Sansone G., Cerullo G., Nisoli M., De Silvestri S., Stagira S., Millijoule-level phase-stabilized few-optical-cycle infrared parametric source, Opt. Lett., Vol. 32, No. 20, pp. 2957-2959 (2007)

Chapter 10

A UNIFIED APPROACH TO ANALYSING THE ANISOPLANATISM OF ADAPTIVE OPTICAL SYSTEMS

Jingyuan Chen[1] and Xiang Chang[1]

[1]Yunnan Astronomical Observatory, Chinese Academy of Science, China

INTRODUCTION

To improve the quality of a laser beam propagating in atmospheric turbulence or to improve the resolution of turbulence-limited optical systems, adaptive optics (AO) (Hardy 1998; Tyson 2011) has been developed. In classical AO systems, the compensation is realized by real-time detection of the turbulence-induced perturbations from a source (beacon) using a wave-front sensing device and then removing them by adding a conjugated item on the same path using a wave-front compensating device.

However, the perturbations caused by the beacon and the target may not be the same, so when the perturbations measured by the beacon are used to compensate the perturbations caused by the target, the compensation performance is degraded. These effects are referred to as anisoplanatism (Sasiela 1992). Anisoplanatic effects are present if there is a spatial separation between the target and beacon (Fried 1982), a spatial separation between the wave-front sensing and compensating apertures (Whiteley, Welsh et al. 1998), when time delays in the system cause the beacon phase and the target phase to be only partially corrected due to atmospheric winds or motion of the system components (Fried 1990) or when the beacon and target have different properties such as distributed size (Fried 1995; Stroud 1996) or wavelength (Wallner 1977), and so on.

Conventionally, all kinds of anisoplanatic effects are studied individually, assuming that they are statistically uncorrelated, and the total effects are obtained by summing them all together when necessary (Gavel, Morris et al. 1994). This conventional approach has a rich history dating back to the earliest days of AO technology and has obtained many good results. But this approach is very limited, because for actual applications of AO systems, many kinds of anisoplanatic effects exist simultaneously and are dependent on each other (Tyler 1994). It is increasingly obvious that these methods are inadequate to treat the diverse nature of new AO applications and the concept of anisoplanatism, and the associated analysis methods must be expanded to treat these new systems so their performance may be properly assessed.

Although anisoplanatism takes many forms, it can be quantified universally by the correlative properties of the turbulence-induced phase. Therefore, instead of investigating a particular form of anisoplanatism, this paper concentrates on constructing a unified approach to analyse general anisoplanatic effects and their effects on the performance of AO systems. For the sake of brevity, we will consider only the case of classic single-conjugate AO systems and not consider the case of a multi-conjugate AO system (Ragazzoni, Le Roux et al. 2005).

In section 2 the most general analysis geometry with two spatially-separated apertures and two spatially-separated sources is introduced. In section 3, we introduce the transverse spectral filtering method which will be used to develop the unified approach for anisoplanatism in this chapter and the general expression of the anisoplanatic wave-front variance will be introduced. In section 4, some special geometries will be analysed. Under these special geometries, the scaling laws and the related characteristic quantities widely used in the AO field, such as Fried's parameter, the Greenwood frequency, the Tyler frequency, the isoplanatic angle, the isokinetic angle, etc., can be reproduced and generalized. In section 5, two specific AO systems will be studied to illustrate the application of the unified approach described in this chapter. One of these systems is an adaptive-optical bi-static lunar laser ranging system and the other is an LGS AO system where, besides the tip-tilt components, the defocus is also corrected by the NGS subsystem. Simple conclusions are drawn in section 6.

GENERAL ANALYSIS GEOMETRY

In the development that follows, we will employ the geometry shown in Figure 1, which is introduced by Whiteley et al. (Whiteley, Roggemann et al. 1998). This geometry shows two apertures, including sensing aperture and compensation aperture, whose position vectors are given by $\vec{r}_s$ and $\vec{r}_c$. Two optical sources, including target and beacon, are located by position vectors $\vec{r}_t$ and $\vec{r}_b$, respectively. The position vectors of the two apertures and the two sources share a fixed coordinate system. A vertical atmospheric turbulence layer, located at altitude z, is also shown in Figure 1.

The projected separation of the aperture centres in this turbulence layer is given by

$$\vec{s}_z = \gamma_z \vec{d} + (\gamma_z - \alpha_z)\vec{r}_s + A_{tcz}\vec{r}_t - A_{bsz}\vec{r}_b$$

$$(1)$$

where $\vec{d} = \vec{r}_c - \vec{r}_s$ is the distance of two apertures, Absz and Atcz are the layer scaling factors given by $A_{bsz} = [z - (\vec{r}_s \cdot \hat{z})]/[(\vec{r}_b - \vec{r}_s) \cdot \hat{z}]$ and $A_{tcz} = [z - (\vec{r}_c \cdot \hat{z})]/[(\vec{r}_t - \vec{r}_c) \cdot \hat{z}]$, while αz and γz are propagating factors of beacon and target, and defined by $\alpha_z = 1 - A_{bsz}$, and $\gamma_z = 1 - A_{tcz}$.

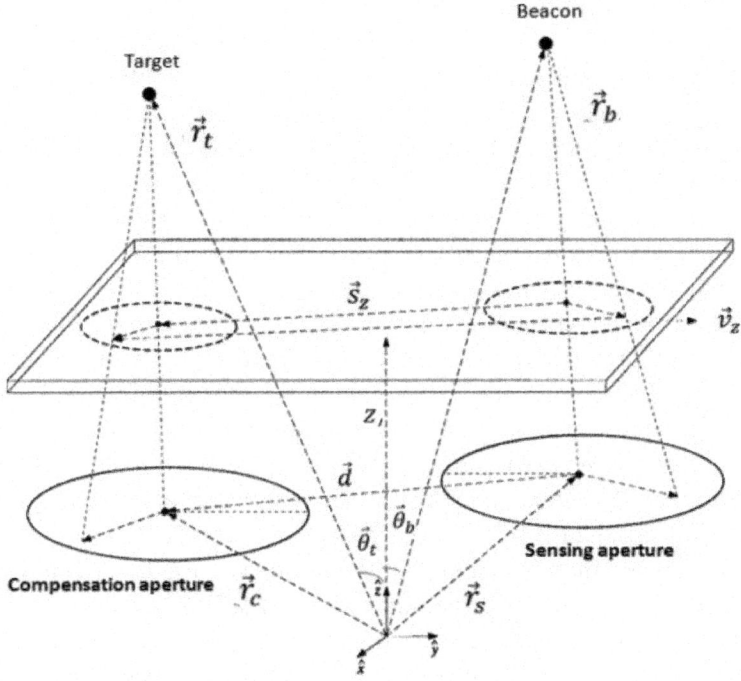

Figure 1: General geometry of the adaptive optical system.

Under some hypotheses, these expressions can be further simplified. We suppose two apertures are at the same altitudes and select the centre of the sensing aperture as the origin of coordinates. We express the positions of target and beacon with the zenith angle and altitude as $(\theta^{\rightarrow} t, L)$ and $(\theta^{\rightarrow} b, H)$, respectively. We notice that in studying anisoplanatic effects, the offsets angular is very small in general (Welsh and Gardner 1991), i.e., $\theta^{\rightarrow} \ll 1$, then Eq. (1) is well approximated by

$$\vec{s}_z = \gamma_z \vec{d} + z\vec{\theta}$$

$$(2)$$

where $\vec{\theta} = \vec{\theta}_t - \vec{\theta}_b$ is the angular separation between target and beacon. At the same time, the propagating factors can be simplified to $\alpha_z = 1 - z/H$, and $\gamma_z = 1 - z/L$.

Further, if we consider delayed-time (τ) of the compensating process, then the projected separation can be expressed as

$$\vec{s}_z = \gamma_z \vec{d} + z\vec{\theta} + \vec{v}_z \tau \tag{3}$$

where $\vec{v}_z$ is the vector of wind velocity in this turbulent layer.

The above is the most general geometric relationship of AO systems. Depending on the conditions of application, more simple geometry can often be used to consider the anisoplanatism of AO systems. Some examples are showed in Figure 2. When the target is sufficiently bright, wave-front perturbation can be measured by directly observing the target. Thus an ideal compensation can be obtained and no anisoplanatism exists. This case is showed in Figure 2(a). In general, the target we are interested in is too dim to provide wave-front sensing, another bright beacon in the vicinity of the target must be used, as depicted in Figure 2(b). In this case, the so-called angular anisoplanatism exits (Fried 1982). In more general cases, a naturally existed object (NGS) cannot be find appropriately, to use AO systems, artificial beacons (LGS) must be created to obtained the wave-front perturbations (Happer, Macdonald et al. 1994; Foy, Migus et al. 1995). Then so called focal anisoplanatism (Buscher, Love et al. 2002;Muller, Michau et al. 2011) appears because of an altitude difference between LGS and target, as depicted in Figure 2(c). Figure 2(d) illustrates that a special anisoplanatism will be induced when a distributed source is used as the AO beacon because it is different from a pure point source (Stroud 1996). Distributed beacons are often occurred, for example, a LGS will wander and expand as a distributed source because of the effects of atmospheric turbulence when the laser is projected upward from the ground (Marc, de Chatellus et al. 2009). In Figure 2(e), the anisoplanatism induced by a separation of the wave-front sensing and compensation aperture is illustrated. With many applications, such as airborne lasers, the separated apertures are indispensable because of the moving platform (Whiteley, Roggemann et al. 1998). Figure 2(f) illustrates a hybrid case, in which many anisoplanatic effects coexist at the same time.

All these special anisoplanatic effects are degenerated cases and can be analysed under general geometry. In the following section, we will construct the general formularies of anisoplanatic variance under the most general geometry.

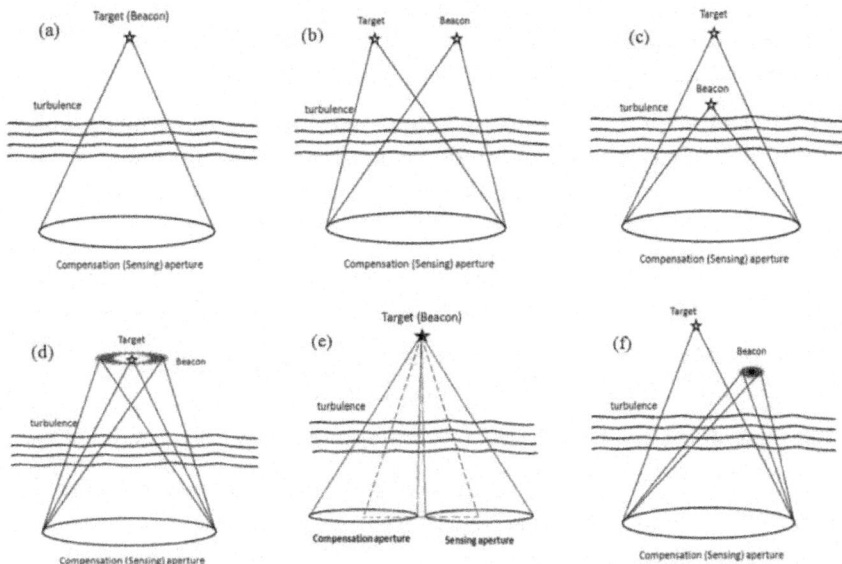

Figure 2: Some special cases of geometry and anisoplanatism. (a) ideal compensation, where the target is also used as the beacon; (b) angular anisoplanatism; (c) focal anisoplanatism; (d) extended beacon; (e) separated apertures; (f) hybrid beacon - many anisoplanatic effects existing at the same time.

TRANSVERSE SPECTRAL FILTERING METHOD AND GENERAL EXPRESSIONS OF CORRECTED (ANISOPLANATIC) WAVE-FRONT VARIANCE

Sasiela and Shelton developed a very effective analytical method to solve the problem of wave propagating in atmospheric turbulence (Sasiela 2007). This method uses Rytov's weak fluctuation theory and the filtering concept in the spatial-frequency domain for coordinates transverse to the propagation direction. In the most general case, the variance of a turbulence-induced phase-related quantity for the propagating waves, when diffraction is ignored, can be written as:

$$\sigma^2 = 2\pi k_0^2 \int_0^L dz (\int \Phi(\vec{\kappa}, z) f(\vec{\kappa}, z) d\vec{\kappa})$$

(4)

where L is the propagation distance and k0 is the space wave number, which when related to wavelength λ by $k_0 = 2\pi / \lambda$; $\Phi(\vec{\kappa}, z)$ is two-dimensional transverse power spectrum of fluctuated refractive-index at the plane vertical to the direction of wave propagation and $\vec{\kappa}=(\kappa,\,\varphi);\,f(\vec{\kappa},\,z)$ is the transverse spectral filter function related to this calculated quantity, whose explicit form can be determined by the corresponding physical processes.

For the atmospheric turbulence, the two-dimensional transverse power spectrum of fluctuated refractive-index can generally be written as:

$$\Phi(\vec{\kappa},\,z)=0.033C_n^2(z)g(\kappa)\kappa^{-11/3}$$

(5)

where $C_n^2(z)$ is refractive-index structure parameter which is allowed to vary along the propagation path, and $g(\kappa)$ is the normalized spectrum. If $g(\kappa)=1$, then the classic Kolmogrov spectrum is obtained.

Now substitute Eq. (5) into Eq. (4), and sequentially perform the integration of wave vector $\vec{\kappa}=(\kappa,\,\varphi)$ at the angular and radial components (Sasiela and Shelton 1993), then the variance reduces to

$$\sigma^2=0.4147\pi k_0^2\!\int_0^L dz C_n^2(z)I_F(z)$$

(6)

in which radial and angular integration can be written respectively as:

$$I_F(z)=\int_0^\infty F(\kappa,\,z)g(\kappa)\kappa^{-8/3}d\kappa$$

(7)

$$F(\kappa,\,z)=\frac{1}{2\pi}\int_0^{2\pi} f(\vec{\kappa},\,z)d\varphi$$

(8)

To evaluate the integral Eq. (6), the expression of the filter function must be given. We will introduce the anisoplanatic filter function for general geometry illustrated in Figure 1. The anisoplanatic filter function can be created from some complex filter functions, describing the process related to the observed target and beacon respectively, by taking the absolute value squared of their difference.

Clearly, when $z\geq H$, the anisoplanatic filter function is

$$f(\vec{\kappa},\,z)=|G(\gamma_z\vec{\kappa})|^2$$

(9)

While when $z<H$, this item can be expressed as:

$$f(\vec{\kappa},\,z)=|G(\gamma_z\vec{\kappa})exp(i\vec{\kappa}\bullet\vec{s}_z)-G(\alpha_z\vec{\kappa})G_s(\vec{\kappa},\,z)|^2$$

(10)

In above two equations, $G(\vec{\kappa})$ is the complex filter function corresponding to the wanted quantity, while $G_s(\vec{\kappa},z)$ is a complex function which can describe the characteristic of the beacon (such as distributed or point-like). When writing this equation, we have supposed that the main physical processes are linear and their complex filter function can be cascaded to form the total filter functions.

Below we list some explicit expressions of complex filter functions.

The transverse complex filter function for a uniform, circular source with angular diameter θr, can be expressed as:

$$G_s(\vec{\kappa}, z) = 2J_1(\kappa\theta_r z)/(\kappa\theta_r z)$$

(11)

Here $J_n(\cdot)$ is the nth-order of Bessel function of the first kind; Similarly, the filter function for a Gaussian intensity distribution with $1/e$ radius θr, has a complex filter function

$$G_s(\vec{\kappa}, z) = exp[-(\kappa\theta_r z)^2/2]$$

(12)

We also notice that for a point-like beacon, the filter function is simply 1.

For the global phase, the complex filter function is

$$G_\phi(\vec{\kappa}, \vec{\rho}) = exp(i\,\vec{\kappa}\bullet\vec{\rho})$$

(13)

The expression of the complex filter function for Zernike mode $Z(m,n)$ depends on its radial (n) and azimuthal (m) order. For $m=0$ it can be written as:

$$G_{n,0}(\vec{\kappa}) = (-1)^{n/2}N_n(\vec{\kappa})$$

(14)

For $m\neq0$ it is given by

$$\left.\begin{matrix} G_{n,m}^x(\vec{\kappa}) \\ G_{n,m}^y \vec{\kappa} \end{matrix}\right\} = i^m\sqrt{2}(-1)^{(n-m)/2}N_n(\vec{\kappa})\begin{cases} cos(m\varphi) \\ sin(m\varphi) \end{cases}$$

(15)

In previous two equations, D is the diameter of aperture and

$$N_n(\vec{\kappa}) = 2\sqrt{n+1}J_{n+1}(\kappa D/2)/(\kappa D/2)$$

(16)

By the above complex filter functions, the expressions of anisoplanatic filter functions of global phase and its Zernike modes can be established explicitly.

For the total phase, When $z\geq H$, from Eq. (9) and Eq. (13), it is

$$F_\phi(\kappa, z) = 1 \tag{17}$$

While when z<H, from Eq. (10) and Eq. (13), the result is

$$F_\phi(\kappa, z) = 1 - 2N_0((\gamma_z - \alpha_z)\kappa)G_s(\kappa, z)J_0(s_z\kappa) + G_s^2(\kappa, z) \tag{18}$$

Similarly, the anisoplanatic filter functions for Zernike modes can also be established. For the case z<H, when m=0, it can be given by the expression

$$F_{n,0}(\kappa, z) = N_n^2(\gamma_z\kappa) + N_n^2(\alpha_z\kappa)G_s^2(\kappa, z) - 2N_n(\gamma_z\kappa)N_n(\alpha_z\kappa)G_s(\kappa, z)J_0(s_z\kappa) \tag{19}$$

When m≠0, for the x, y component of Zernike mode, we can write their anisoplanatic filter functions as follows:

$$F_{n,m}^x(\kappa, z) = N_n^2(\gamma_z\kappa) + N_n^2(\alpha_z\kappa)G_s^2(\kappa, z) + $$
$$-2N_n(\gamma_z\kappa)N_n(\alpha_z\kappa)G_s(\kappa, z)[\, J_0(s_z\kappa) - (-1)^m J_{2m}(s_z\kappa)] \tag{20}$$

$$F_{n,m}^x(\kappa, z) = N_n^2(\gamma_z\kappa) + N_n^2(\alpha_z\kappa)G_s^2(\kappa, z)$$
$$-2N_n(\gamma_z\kappa)N_n(\alpha_z\kappa)G_s(\kappa, z)[\, J_0(s_z\kappa) + (-1)^m J_{2m}(s_z\kappa)] \tag{21}$$

It is easy to find that if we define a new quantity as follows:

$$F_{n,m}(\kappa, z) = F_{n,m}^x(\kappa, z) + F_{n,m}^y(\kappa, z) \tag{22}$$

then we can obtain

$$F_{n,m}(\kappa, z) = C_m[N_n^2(\gamma_z\kappa) + N_n^2(\alpha_z\kappa)\,G_s^2(\kappa, z) - 2\,N_n(\gamma_z\kappa)\,N_n(\alpha_z\kappa)\,G_s(\kappa, z)\,J_0(s_z\,\kappa)] \tag{23}$$

Where Cm is a constant factor related to the azimuthal order mm. If m=0, then Cm=1; otherwise Cm=2.

Similarly for the case z≥H, the corresponding result is

$$F_{n,m}(\kappa, z) = C_m N_n^2(\gamma_z\kappa) \tag{24}$$

SOME SPECIAL CASES

In the previous section the transverse anisoplanatic spectral filter functions for the general geometry of adaptive optical systems have been established. In this section we consider some special geometric cases, where asymptotic solutions of integrals can be obtained.

The Anisoplanatism Induced By Separated Apertures And Its Related Characteristic Distances

We first consider a simple case, where only the anisoplanatism induced by two separated apertures exists and the others are ignored. Let $\tau = 0$, $\theta = 0$, $G_s(\kappa, z) = 1$, and $L = H = +\infty$ (i.e., $\gamma_z = \alpha_z = 1$), and taking into account the limitation $\lim_{x \to 0} N_0(x) = 1$, then Eq. (18) and Eq. (23) reduce to

$$F_\phi(\kappa, z) = 2[1 - J_0(\kappa d)]$$

(25)

$$F_{n,m}(\kappa, z) = 2C_m N_n^2(\kappa)[1 - J_0(\kappa d)]$$

(26)

The anisoplanatic phase variance is easily obtained. Substituting Eq. (25) into Eq. (6), and using the Kolmogrov spectrum, i.e., $g(\kappa) = 1$ the integral is equal to

$$\sigma_\phi^2 = (d / d_0)^{5/3}$$

(27)

Here we have calibrated the variance with a new characteristic distance d0 defined as

$$d_0 = 0.526 \, k_0^{-6/5} \mu_0^{-3/5}$$

(28)

This is about 1/3 of the atmospheric coherence length $r_0 = (0.423 k_0^2 \mu_0)^{-3/5}$, Where μ_m represents the mth (full) turbulence moments. From Eq. (27), we find that the anisoplanatic variance induced by separated apertures meets the 5/3 power scaling law with the distance of separated apertures.

For AO systems, the piston phase variance is not meaningful and can be removed from the total variance. Their difference, i.e., the piston-removed phase variance, cannot be expressed analytically for arbitrary distances, while for very small and very large distance their asymptotic solutions can be found. We first calculate in these limitations the wave vector integral of the piston-removed anisoplanatic phase filter function $I_{\phi eff,aniso} = I_\phi - I_{0'}$ which can be found easily from the Eq. (79) and (80) in Appendix with n=0.

When d≫D expanding $I_{\phi eff,aniso}$ to second order of (D/d), the result is

$$I_{\phi eff,aniso} \sim \left(\frac{D}{2}\right)^{5/3}\left[-\frac{4\,\Gamma(-5/6)\Gamma(7/3)}{\sqrt{\pi}\,\Gamma(17/6)\Gamma(23/6)} - \frac{\Gamma(1/6)}{4\,\Gamma(5/6)}\left(\frac{D}{d}\right)^{1/3}\right]$$

(29)

While d≪D, expanding it to fourth-order of (d/D) the result is

$$I_{\phi eff,aniso} \sim -\frac{\Gamma(-5/6)}{\Gamma(11/6)}\left(\frac{D}{2}\right)^{5/3}\left(\frac{d}{D}\right)^{5/3} + \frac{\Gamma(7/3)\Gamma(-5/6)}{\sqrt{\pi}\,\Gamma(17/6)\Gamma(23/6)}\left(\frac{D}{2}\right)^{5/3}\left(\frac{d}{D}\right)^2\left[\frac{51425}{41472}\left(\frac{d}{D}\right)^2 - \frac{935}{144}\right]$$

(30)

On the other hand, the wave vector integral of the piston-removed phase filter function for a single wave beam is easy to find and can be expressed as:

$$I_{\phi eff,single} = -\frac{2\,\Gamma(-5/6)\Gamma(7/3)}{\sqrt{\pi}\,\Gamma(17/6)\Gamma(23/6)}\left(\frac{D}{2}\right)^{5/3}$$

(31)

From the above equations we find that in the limitation of d≫D the piston-removed anisoplanatic phase variance tends to be twice that of the piston-removed phase variance of a single wave. This is predictable, because when the separated distance of apertures is large enough, the correlation of waves from two separated aperture is gradually lost, and these beams are statistically independent of each other. We also find in the limitation of d≪D the piston-removed anisoplanatic phase variance remains the 5/3 power scaling law with the separated distance, which is same as that for the total phase in Eq. (27).

There are many ways to define a related characteristic distance. For an AO system, if the piston-removed anisoplanatic phase variance is greater than the same quantity for a single wave, that is to say

$$\sigma^2_{\phi eff,aniso}\Big|_{d \ll D} > \sigma^2_{\phi eff,single}$$

(32)

Then the compensation is ineffective and the AO system is not needed. We can define the uncorrected distance dunc of two separated apertures as the smallest distance satisfied above inequality. Using Eq. (30), Eq. (31) and Eq. (32), an approximation of this characteristic distance can be given by dunc=0.828D.

On the other hand, to achieve a better performance, the residual error of corrected wave must be small enough. Similar to the isoplanatic angle, we can define the isoplanatic distance as the separated distance of apertures at which the residual error is an exact unit. From the scaling law of Eq. (27), this distance is same as d0d0, i.e., diso=d0.

The above two characteristic distances (dunc and diso) give different restrictions to an apertures-separated AO system. Other characteristic distances can also be defined. For example, for such an AO system, we can define the effective corrected distance (deff) as the separated distance of apertures at which the AO system can work effectively. Obviously this distance can be determined by the smaller of the above two characteristic distances, namely,

$$d_{eff} = Min\{d_{iso}, d_{unc}\}$$

(33)

In general, the inequality $d_{iso} < d_{unc}$ is always satisfied, so the effective corrected distance is $d_{eff} = d_{iso} = d_0$.

Similar to the above analysis and definitions for total phase, anisoplanatic variances and related characteristic distances can be determined for arbitrary Zernike modes. The final result is complex and can be expressed with generalized hypergeometric functions (Andrews 1998). In order to obtain a simpler close solution, we consider the limit case of very large or very small separating distance.

From Eq. (80), in the limitation d≪D the integral is approximately equal to

$$I_{n,m}(z) = C_m \frac{11 \, \Gamma(7/3)\Gamma(n + 1/6)(1 + n)}{2^{8/3}\sqrt{\pi} \, \Gamma(17/6)\Gamma(17/6 + n)} \left(\frac{d}{D}\right)^2 D^{5/3}$$

(34)

Furthermore, performing the integration at the propagating path, the asymptotic value of the anisoplanatic phase variance for Zernike mode Z(m,n) is obtained as follows:

$$\sigma_{n,m}^2 = 0.879 C_m k_0^2 \frac{(1 + n)\Gamma(n + 1/6)}{\Gamma(n + 17/6)D^{1/3}} \mu_0 d^2$$

(35)

If we defined the isoplanatic distance of the Zernike mode $Z(m,n)$ $d_{n,m;iso}$ as the distance satisfied the condition $\sigma_{n,m}^2 = 1$, then the variance can be calibrated as:

$$\sigma_{n,m}^2 = (d / d_{n,m;iso})^2$$

(36)

This characteristic distance can be determined as follows:

$$d_{n,m;iso} = \frac{1.067}{k_0} \sqrt{\frac{\Gamma(n + 17/6) \, D^{1/3}}{C_m \mu_0 (n + 1)\Gamma(n + 1/6)}}$$

(37)

For a single beam, the expression corresponding to Eq. (34) is (n≥1)

$$I_{n,m;single}(z) = \frac{\Gamma(7/3)\Gamma(n - 5/6)(1 + n) \, C_m}{2^{2/3}\sqrt{\pi} \, \Gamma(17/6)\Gamma(n + 23/6)} D^{5/3}$$

(38)

From Eq. (34) and Eq. (37), and another inequality similar to Eq. (32), the uncorrected distance of the Zernike mode Z(m,n) can be defined as:

$$d_{n,m;unc} = 12D / \sqrt{11(6n-5)(6n+17)}$$

(39)

Similarly, the effective distance of the Zernike mode $Z(m,n)$ can be defined as (at $n \geq 1$):

$$d_{n;eff} = \underset{m}{Min} \left(d_{n,m;iso}, d_{n,m;unc} \right)$$

(40)

When the separated distance of the two apertures is smaller than this characteristic distance, the Zernike mode $Z(m,n)$ of turbulence-induced phase can be compensated effectively by the AO system. In Eq. (40), the Minimum operator is evaluated throughout all the field of m, so the result is no longer dependent on m.

In Figure 3, we show the typical values of the characteristic distances $d_{n;eff}$' defined above for the separated-apertures-induced anisoplanatism with D=1.2m. As a comparison with the total phase, the value of piston-removed quantity d0d0 is also showed in the same figures at n = 0.

In Figure 3(a), the relationship among $d_{n;eff}$' and the other two characteristic distances (for $d_{n,m;iso}$ and $d_{n,m;unc}$', their values also select the minimum in all the ms) are showed for λ=532nm. From this figure, we find that the isoplanatic distance is monotonous - increasing with the radial order of Zernike mode - while the uncorrected distance is decreasing with it. Therefore, the effective distance is determined by the isoplanatic distance when the radial order is small (such as for the tip-tilt, defocus, et al) and by the uncorrected distance when the radial order is large. We also find that the effective distances for small ns are usually greater than those for the (piston-removed) total phase, so when only a few low-level Zernike modes need to be compensated for, apertures with greater separated distance can be used.

Other sub-figures in Figure 3 show the effective distances for different compensational orders at different turbulent intensities and wavelengths. In Figure 3(b), four different turbulent intensities (r0=3cm, 6cm, 9cm and 12cm at reference wavelength of λ=500nm) are compared. In Figure 3(c), the effective distances for two different turbulence intensities (r0=5cm and 10cm) and two different wavelengths (λ=532nm and 1064nm) are compared. We can find that the effective distances are smaller at stronger turbulences or smaller wavelengths.

In Figure 3(d), the relationships between the effective distances and turbulence intensities are showed for four different compensational orders (n=1, 2, 3, and 5) at λ=532nm. This shows that the effective (or uncorrected) distances are not related to the turbulence intensities for lager compensational orders, such as that for n=5.

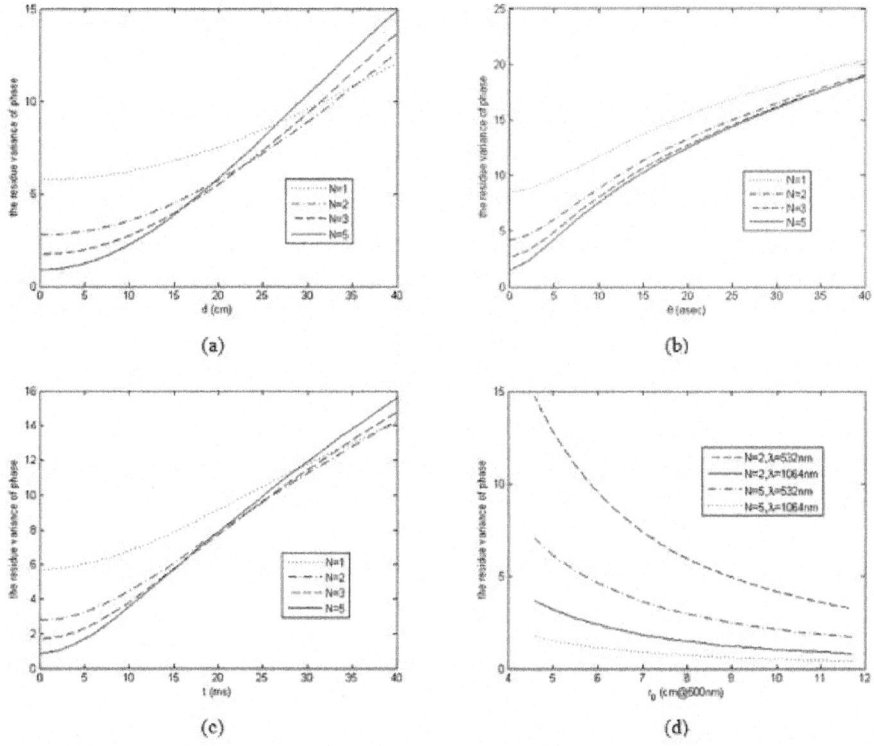

(a)

(b)

(c)

(d)

Figure 3: The characteristic distances for the anisoplanatism of separated apertures. (a) The relationship among three characteristic distances, λ=532nm; (b) The effective distances at for four different turbulent intensities, λ=532nm; (c) The effective compensational distances at different turbulent intensities and wavelengths; (d) The relationship between the effective distances and turbulence intensities for four different compensational orders, λ=532nm.

THE ANGLULAR ANISOPLANATISM AND RELATED CHARACTERISTIC ANGLES

Now we consider the geometry where only angular anisoplanatism exits. Let $d=0\tau=0$ $\gamma_z=\alpha_z=1$, and $G_s(\kappa, z)=1$, then Eq. (18) and Eq. (23) reduce to

$$F_\phi(\kappa, z)=2[1 - J_0(\kappa\theta z)]$$

(41)

$$F_{n,m}(\kappa, z)=2C_m N_n^2(\kappa)[1 - J_0(\kappa\theta z)]$$

(42)

Substituting Eq. (41) into Eq. (6), and using the Kolmogrov spectrum, the result is

$\sigma_{\varphi}^2 = (\theta / \theta_0)^{5/3}$, here θ_0 is the well-known isoplanatic angle defined as (Fried 1982) $\theta_0 = (2.914 k_0^2 \mu_{5/3})^{-3/5}$.

Similarly, in the limitation of very small offset angle, i.e., $\theta z \ll D$, the effective corrected offset angle between beacon and target can be defined and determined by $\theta_{eff} = \theta_{0'}$.

Using Eq. (42), the angular anisoplanatism of Zernike modes can also be calculated. The results can be expressed with the generalized hypergeometric functions, and in some limit conditions, a more compact expression can be obtained.

We consider the limitation of $\theta z \ll D$. Using Eq. (80) in Appendix, the angular anisoplanatism of Zernike mode $Z(m,n)$ can be expanded to the turbulence second-order structure constant moments and can be expressed as ($n \geq 1$)

$$\sigma_{n,m}^2 = (\theta / \theta_{n,m;iso})^2$$

(43)

where the characteristic angle

$$\theta_{n,m;iso} = \frac{1.06665}{k_0 \mu_2^{1/2}} \sqrt{\frac{\Gamma(n + 17/6) D^{1/3}}{C_m (n + 1) \Gamma(n + 1/6)}}$$

(44)

can be defined as the is oplanatic angle for Zernike mode $Z(m,n)$, and it is the size of the offset-axis angle between the beacon and the target when the angular anisoplanatism of Zernike mode is unit rad^2.

When n=1and m=1, the tip-tilt isoplanatic angle (also called isokinetic angle) is obtained. This characteristic angle can be expressed as:

$$\theta_{TA} = \theta_{1,1;iso} = 1.224 (k_0^2 \mu_2 D^{-1/3})^{-1/2}$$

(45)

This is consistent with the results in other research (Sasiela and Shelton 1993).

Similar to anisoplanatism of separated apertures, other characteristic angles can be defined and calculated. The uncorrected offset angle of $Z(m,n)$ can be expressed as:

$$\theta_{n,m;unc} = 12 D \sqrt{\mu_0 / \mu_2} / \sqrt{11(6n - 5)(6n + 17)}$$

(46)

and the effective offset angle of the n-order Zernike mode can be determined by

$$\theta_{n;eff} = \min_{m} \left(\theta_{n,m;iso}, \theta_{n,m;unc} \right)$$

(47)

In Figure 4, the typical values of the characteristic angles $\theta_{n;eff}$ defined above are showed at D=1.2m. In Figure 4(a), we compare the values for two different turbulent intensities (r_0=5cm, 10cm) and two different wavelengths (λ=532nm, and 1064nm). We can also find that the effective offset angles $\theta_{n;eff}$ small ns are usually greater than those for the (piston-removed) total phase, the same as the characteristic quantities $d_{n;eff}$. In fact, this is one of main reasons that the use of LGS can partially solve the so-called "beacon difficulty", because a NGS may be find to correct the lower order modes of the turbulence-induced phase in a field far wider than that limited by the isoplanatic angle θ_0. Unlike $d_{n;eff}$, the effective offset angle $\theta_{n;eff}$ is not only dependent on aperture diameter D, but also turbulence intensity. Therefore, for higher-order Zernike modes, the effectively offset angle is also dependent on the turbulence intensity. In Figure 4(b), the relationships between effective offset angles and turbulence intensities are showed for four different compensational orders (n=1, 2, 3, and 5) atλ=532nm.

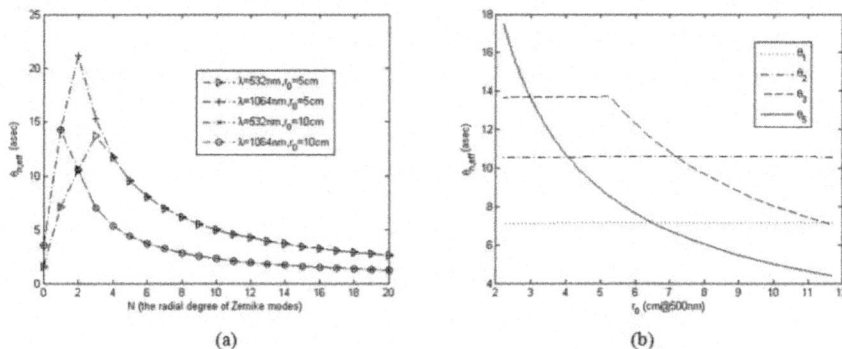

(a)   (b)

FIGURE 4: The characteristic angles of the angular anisoplanatism for separated beacon and target. (a) the effective offset angles at different turbulent intensities and wavelengths; (b) the relationship between effective offset angles and turbulence intensities for four different compensational orders atλ=532nm.

The Time-Delayed Anisoplanatism And Related Characteristic Quantities

When $d=0$, $\theta=0$, $\gamma_z = \alpha_z = 1$, and $G_s(\kappa, z)=1$, and $G_s(\kappa,z)=1$, then there is only time-delayed anisoplanatism. Now Eq. (18) and Eq. (23) reduce to

$$F_\phi(\kappa, z) = 2[1 - J_0(\kappa\, v_z \tau)] \tag{48}$$

$$F_{n,m}(\kappa, z) = 2C_m N_n^2(\kappa)[1 - J_0(\kappa\, v_z \tau)] \tag{49}$$

Using Eq. (48) and $g(\kappa) = 1$ to perform the integration in Eq. (6), the total phase anisoplanatism variance can be expressed as $\sigma_\phi^2 = (\tau/\tau_0)^{5/3}$, where τ_0 is normally-defined atmospheric coherence time and equal to $\tau_0 = (2.913 k_0^2 v_{5/3})^{-3/5}$, and v_n is the nwth velocity moments of atmospheric turbulence defined by $v_n = \int_0^L dz C_n^2(z) v^m(z)$. This characteristic quantity $\tau 0$ is related to the Greenwood frequency. For a single-poles filter (controller), the variance of compensated phase can be scaled as $\sigma_\phi^2 = (f_0/f_{3db})^{5/3}$, , where f is the effective control bandwidth of AO system and f0 is the Greenwood frequency, defined by $f_0 = (0.103 k_0^2 v_{5/3})^{3/5}$. We can easily find there is a simple relationship between these two characteristic quantities:

$$f_0 \approx 0.134/\tau_0 \tag{50}$$

as is first noted by Fried (Fried 1990).

Similarly, in the limitation $v_z\tau \ll D$, the effective corrected time can be defined and determined by $\tau_{eff} = \tau 0$. For arbitrary Zernike mode of phase, from Eq. (80), when we consider the second order approximation, the anisoplanatic variance is equal to

$$\sigma_{n,m}^2 = 0.879 \frac{(1+n)\Gamma(n+1/6)}{\Gamma(n+17/6)} C_m k_0^2 D^{-1/3} v_2 \tau^2 \tag{51}$$

Using the isoplanatic time $\tau_{n,m;iso}$ satisfied $\sigma_{n,m}^2 = 1$ to rescale, then the variance can be expressed as:

$$\sigma_{n,m}^2 = (\tau/\tau_{n,m;iso})^2 \tag{52}$$

and its expression is

$$\tau_{n,m;iso} = \frac{1.06665}{k_0 v_2^{1/2}} \sqrt{\frac{\Gamma(n+17/6)\, D^{1/3}}{C_m(n+1)\Gamma(n+1/6)}} \tag{53}$$

Similar to Greenwood frequency, we can apply Eq. (50) to define a characteristic frequency related to the isoplanatic time in Eq. (53) as follows:

$$f_{n,m;iso} = 0.1256 k_0 v_2^{1/2} \sqrt{\frac{C_m(n+1)\Gamma(n+1/6)}{\Gamma(n+17/6)\, D^{1/3}}} \tag{54}$$

This is the characteristic frequency using a single-poles filter to compensate for the Zernike mode $Z(m,n)$ of the turbulence-induced phase.

Further, the effective correction time of arbitrary n-order Zernike model of phase can be defined as:

$$\tau_{n;eff} = \min_{m} \left(\tau_{n,m;iso}, \tau_{n,m;unc} \right) \tag{55}$$

where the uncorrected time can be expressed as

$$\tau_{n,m;unc} = 12 \, D\sqrt{\mu_0 / v_2} / \sqrt{11(6n - 5)(6n + 17)} \tag{56}$$

When using an AO system with a time delay exceeding this characteristic time to compensate for the n-order Zernike model of phase, the compensation is ineffective.

The characteristic quantities $\tau_{n;eff}$ are similar to $\theta_{n;eff}$ and $d_{n;eff}$. In Figure 5, we show some typical values of the characteristic angles $\tau n;eff$.

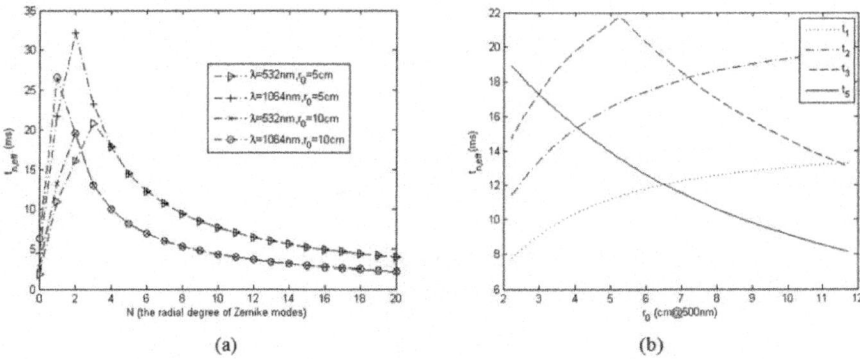

(a)   (b)

Figure 5: The characteristic times for the time-delay anisoplanatism. (a) The effective times at different turbulent intensities and wavelengths; (b) The relationship between effective times and turbulence intensities for different compensational orders at $\lambda = 532nm$.

When $n = 1$ and $m = 1$, the isoplanatic times or the characteristic frequencys for the tip-tilt component of the turbulence-induced phase are obtained as:

$\tau_{1,1;iso} = (0.668k_0^2 v_2 D^{-1/3})^{-1/2}$ $f_{1,1;iso} = 0.4864 \, \lambda^{-1} v_2^{1/2} D^{-1/6}$. It should be noted that these results are slightly different with others. In many studies, the tilt anisoplanatic variances are calibrated as: $\sigma_\alpha^2 = \frac{1}{5}\left(\frac{\lambda}{D}\right)^2\left(\frac{\tau_{st}}{\tau_{0t}}\right)^2$ or $\sigma_\alpha^2 = \left(\frac{\lambda}{D}\right)^2\left(\frac{f_T}{f_{3db}}\right)^2$, Where the characteristics time (Parenti and Sasiela 1994) and frequency (Tyler 1994) are defined by $\tau_{0t} = \left(0.512 k_0^2 v_{-1/3}^{8/15} v_{14/3}^{7/15} D^{-1/3}\right)^{-1/2}$ or $f_T = 0.368 \lambda^{-1} v_2^{1/2} D^{-1/6}$. These results are slightly different from ours because different methods of series expanding are used. However, the differences are minor and our expressions have simpler forms and are more convenient to use.

The Focal Anisoplanatism

If the altitudes of beacon and target are different, then focal anisoplanatism appears. When other anisoplanatic effects are neglect (i.e., $\theta = 0$, $d = 0$, $\tau = 0$, $L = +\infty$, $G_s(\kappa, z) = 1$), the anisoplanatic filter function below the beacon are simplified to

$$F_\phi(\kappa, z) = 2\left[1 - \frac{H}{\kappa D z} J_1\left(\frac{\kappa D z}{2H}\right)\right]$$

(57)

$$F_{n,m}(\kappa, z) = C_m[N_n(\kappa) - N_n(\alpha_2 \kappa)]^2$$

(58)

Substituting Eq. (57) into Eq. (6), the anisoplanatic variance for total phase is given by $\sigma_\phi^2 = 0.5 k_0^2 \mu_{5/3}^- (D/H)^{5/3}$, here μ_m^- is the mth lower turbulence moment, defined by $\mu_m^- = \int_0^H C_n^2(z) z^m dz$.

Similarly, using Eq. (58) the anisoplanatic variance of Zernike mode $Z(m,n)$ can also be calculated. In order to obtain a more simple close solution, we consider the limit case of a very high altitude beacon, i.e., $H \gg z$. From Eq. (81), when the second-order small quantities are retained, the anisoplanatic variance for $Z(m,n)$ can be approximated by

$$\sigma_{n,m}^2 = \frac{3.317}{6237} \frac{(1+n)\Gamma^2(-8/3)[108n(n+2) - 55]}{\Gamma(-10/3)\Gamma(n+23/6)/\Gamma(n-5/6)} C_m k_0^2 \mu_2^- D^{5/3} / H^2$$

(59)

By this expression, the first two components, i.e., the anisoplanatic variances of the piston and tip-tilt, can be obtained immediately as follows:

$\sigma_P^2 = \sigma_{0,0}^2 = 0.0834 k_0^2 D^{5/3} \mu_2^- / H^2$ and $\sigma_T^2 = \sigma_{1,1}^2 = 0.3549 k_0^2 D^{5/3} \mu_2^- / H^2$.

When analyzing a LGS AO system with a telescope aperture of diameter D, it is useful to express the anisoplanatic variance by $\sigma_\phi^2 = (D/d_e)^{5/3}$, where the characteristic quantity de is a measure of effective diameter of the LGS AO system (Tyler 1994) (i.e., a telescope with a diameter equal to dewill have 1 rad of rms wave-front error). Considering the fact that for a LGS system piston is meaningless and tip-tilt is non-detectable (Rigaut and Gendron 1992; Esposito, Ragazzoni et al. 2000), then an approximated value of de can be obtained by

$$d_e = \left\{ k_0^2 \left[0.5 \mu_{5/3}^- / H^{5/3} - 0.4383 \mu_2^- / H^2 \right] \right\}^{-3/5}$$

(60)

We can further consider the effect of turbulence above the beacon. From Eq. (17) and Eq. (24), the filter functions for the total phase and its Zernike mode Z(m,n) above the beacon are

$$F_\phi(\kappa, z) = 1$$

(61)

$$F_{n,m}(\kappa, z) = C_m N_n^2(\kappa)$$

(62)

Therefore the anisoplanatic filter function of the partial phase in which the components of the piston and tip-tilt are removed can be expressed as:

$$F_{eff,up}(\kappa, z) = F_\phi(\kappa, z) - F_{0,0}(\kappa, z) - F_{1,1}(\kappa, z)$$

(63)

Performing the integration Eq. (6), the corresponding variance is obtained as

$$\sigma_{eff,up}^2 = 0.0569 k_0^2 \mu_0^+ D^{5/3}$$

(64)

Where μ_0^+ is the mth upper turbulence moment, defined by $\mu_m^+ = \int_H^\infty C_n^2(z) z^m dz$. So when consider the effect of turbulence above the beacon, the effective diameter can be expressed approximately as

$$d_e = \left\{ k_0^2 \left[0.0569 \mu_0^+ + 0.5 \mu_{5/3}^- / H^{5/3} - 0.4383 \mu_2^- / H^2 \right] \right\}^{-3/5}$$

(65)

This is the same result as that obtained in other research (Sasiela 1994).

The Anisoplanatism Induce By An Extended Beacon

We now consider the anisoplanatic effect induced by a distributed beacon and neglect other anisoplanatic effects. Let $d=0$, $\theta=0$, $\tau=0$, and $\gamma_z=\alpha_z=1$, then Eq. (18) and Eq. (23) are reduced to

$$F_\phi(\kappa, z) = [1 - G_s(\kappa, z)]^2$$

(66)

$$F_{n,m}(\kappa, z) = C_m N_n^2(\kappa)[1 - G_s(\kappa, z)]^2$$

(67)

Substituting above two equations into Eq. (6) and performing the integration, the anisoplanatic variance of the total phase and its Zernike components can be obtained. Below we give the corresponding results for a

Gaussian distributed beacon and Kolmogrov's turbulent spectrum, i.e., using Eq. (12) and $g(\kappa)=1$.

For the total phase, the integration can easily be obtained. The result is $\sigma_\phi^2 = 0.5327 \, \theta_r^{5/3} \mu_{5/3} = (0.3608 \, \theta_r / \theta_0)^{5/3}$, here μ_m here µm is the mth turbulence moment, and θ0 is atmospheric isoplanatic angle. Obviously, the result is similar to the classic 5/3 power scaling law for angular anisoplanatism.

For Zernike component Z(m,n), we consider the limit case of very big θr, i.e., θrz ≫ D. From Eq. (82), the approximate results expanding to the second order turbulence moment can be obtained. Here, we only list the first two components (i.e., the anisoplanatic variances of piston and tip-tilt) as follows: $\sigma_P^2 = \sigma_{0,0}^2 = 0.5327 \theta_r^{5/3} \mu_{5/3} - 0.4369 \, D^{5/3} \mu_0$ .and $\sigma_T^2 = \sigma_{1,1}^2 = 0.3799 \, D^{5/3} \mu_0$.

Two examples for hybrid anisoplanatism

To illustrate the application of the unified approach described in this chapter, we will study two special AO systems as examples in this section. In these examples many anisoplanatic effects exist at the same time, so no analytical solution for anisoplanatic variances can be obtained - only numeric results.

To calculate the anisoplanatic variances, we use the Hufnagel-Valley model:

$$C_n^2(z) = A \, exp\left(-\frac{z}{10^2}\right) + \frac{2.7}{10^{16}} exp\left(-\frac{z}{1500}\right) + \frac{5.94}{10^3}\left(\frac{w}{27}\right)^2\left(\frac{z}{10^5}\right)^{10} exp\left(-\frac{z}{10^3}\right)$$

(68)

where w is the pseudo-wind, and the altitude z expressed in meters. The turbulence strength is usually changed by a variation of the w term or A, the parameter to describe the turbulence strength at the ground. At the same time, the modified von Karman spectrum

$$g(\kappa) = [1 + (\kappa_o/\kappa)^2]^{-11/6} exp[-(\kappa/\kappa_i)^2]$$

(69)

will be use. Where κ_o and κ_i are the space wave numbers corresponding to the outer scale and the inner scale of the atmospheric turbulence, respectively. To consider the effect of time-delay, we use the Bufton wind model

$$v_z = v_g + 30 \, exp[-(z - 9400)^2 / 4800^2]$$

(70)

Where vgis the wind speed on the ground.

An Adaptive-Optical Bi-Static Lunar Laser Ranging (Llr) System

Although the technique of Lunar Laser Ranging (LLR) is one of most important methods to modern astronomy and Earth science, it is also a very difficult task to develop a successful LLR system (Dickey, Bender et al. 1994). One of the main reasons is that the quality of the outgoing laser beams deteriorates sharply due to the effect of atmospheric turbulence, including the wandering, expansion, and scintillation. To mitigate these effects of atmospheric turbulence and improve the quality of laser beams, one can use AO systems to compensate the outgoing beams (Wilson 1994; Riepl, Schluter et al. 1999). In this section we will study the anisoplanatism of a special adaptive optical bi-static LLR system in which the receiving aperture is also used to measure the turbulence-induced wave-front and the outgoing beam is compensated by the conjugated wave-front measured by this aperture. It is a concrete application of the unified approach described in this paper.

For this special AO system, two apertures and the useful point-like beacon (Aldrin, Collins, et al.) and the targets (Apollo_11, Apollo 15, et al.) are separated, so the anisoplanatism is hybrid. Let $G_s(\kappa, z)=1$, $L = H$ $(=3.8 \times 10^8 m)$, , and denote respectively the offset distance and angle of apertures and sources as d and θ then the anisoplanatic filter function in altitude z are reduced to

$$F_\phi(\kappa, z)=2[1 - J_0(s_z\kappa)]$$

(71)

$$F_{n,m}(\kappa, z)=2C_m N_n^2(\gamma_z\kappa)[1 - J_0(s_z\kappa)]$$

(72)

where $\gamma_z=1 - z/L$, $s_z=\gamma_z d + z\theta + v_z\tau$, , and the corrected time delay has been considered. Using above Equations, the variances can be computed easily, but the results can be expressed by higher transcendental functions with no simpler expressions existing.

In Figure 6, we show the anisoplanatic variances when turbulence-induced wave-fronts are compensated to different Zernike orders.

In the first three sub-graphs, the relationships between the anisoplanatic variances and some important parameters (separation distance of apertures, offset angle of sources, time-delay of the correcting process) are also showed respectively. From Figure 6(a), we can see the variances usually monotonously increase with the separated distance. We can also see that increasing the corrected order the variance will decrease when the separated distance is small, but it will not decrease when the separated distance is increased to a certain scale. This is because the effective distances dn;effdn;eff are smaller at larger orders, as has been showed in Figure 3. A similar conclusion can be drawn for the offset angle of sources from Figure 6(b) and for the time delay of the correcting process from Figure 6(c).

In Figure 6 (d), the relationship between anisoplanatic variance and turbulence intensity are showed for two wavelengths (λ=532nm and 1064nm) and two corrected orders (n=2 and 5). In this case, all three anisoplanatic effects (angular, time-delayed and that induced by separated apertures) exist at the same time and the corresponding parameters are selected as $d = 5cm$, $\theta = 2''$, $\tau = 2ms$.

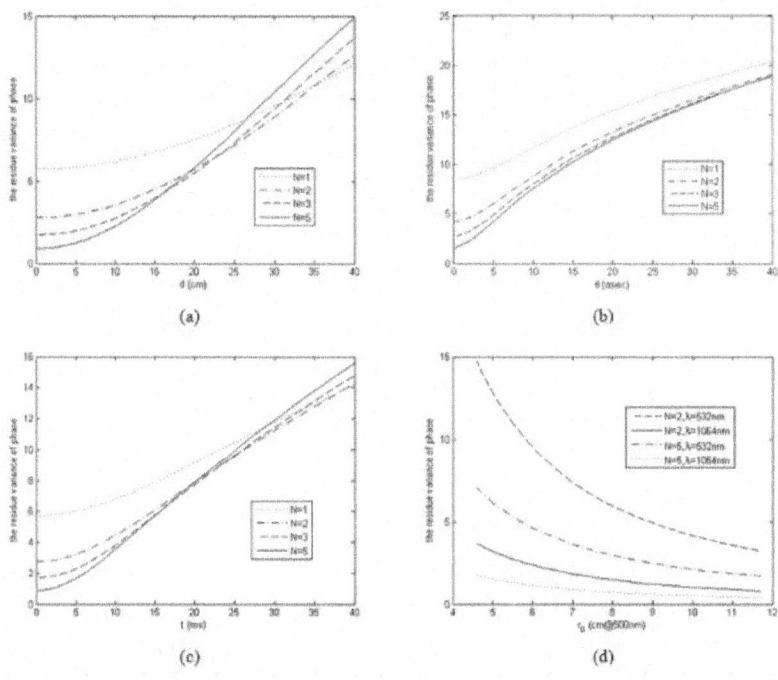

(a)

(b)

(c)

(d)

Figure 6: The anisoplanatic variances for LLR AO system, D=1.2m. (a) The relationship between residual phase variance and separated distances for four different compensational orders at λ=532nm and r0r0=10cm; (b) the relationship between residual phase variance and offset angles; (c) the relationship between residual phase variance and corrected time-delays for four different compensational orders at λ=532nm and r0r0=10cm; (d) the relationship between residual phase variance and turbulent intensities for two different compensational orders (n=2 and 5) and two different wavelengths (λ=532nm and 1064nm).

A Special Lgs Ao System: Defocus Corrected By The Ngs Subsystem

A laser beacon is insensitive to full-aperture tilt because the beam wanders on both the upward and the downward trips through the atmosphere, so currently when using LGS AO systems other NGS subsystems are usually used to sense and correct wave-front tilt. All other Zernike modes except tip-tilt can be corrected by LGS subsystems, but the corrected performance is limited by the focal anisoplanatism. Besides tip-tilt, the defocus (or focus) mode is another main component of the turbulence-induced phase and decreasing the focal anisoplanatism of the defocus component is very important (Esposito, Riccardi et al. 1996; Neyman 1996). In this section, we consider the performance of a special kind of LGS AO system, in which, besides the overall tilt, the focus mode can also be sensed and corrected by the NGS subsystems. Using this special LGS AO system, the focal anisoplanatism of the defocus mode can be reduced further.

We concentrate on the relationship between the focal and angular anisoplanatism of the defocus mode, and neglect the effects induced by time-delay and separated aperture. We also neglect the correlation between LGS and NGS subsystem, and suppose them to be statistically independent of each other. Then the anisoplanatic filter functions for the NGS subsystem are reduced to

$$F_{\phi,N}(\kappa, z) = 1 - 2G_{s,N}(\kappa, z)J_0(\kappa z\theta_N) + G_{s,N}^2(\kappa, z)$$

(73)

$$F_{n,m;N}(\kappa, z) = C_m N_n^2(\gamma_z\kappa)[1 - 2G_{s,N}(\kappa, z)J_0(\kappa z\theta_N) + G_{s,N}^2(\kappa, z)]$$

(74)

While for the LGS subsystem, under the LGS beacon, the results reduce to

$$F_{\phi,L}(\kappa, z) = 1 - 2N_0((\gamma_z - \alpha_z)\kappa)G_{s,L}(\kappa, z)J_0(\kappa z\theta_L) + G_{s,L}^2(\kappa, z)$$

(75)

$$F_{n,m;L}(\kappa, z) = C_m[N_n^2(\gamma_z\kappa) + N_n^2(\alpha_z\kappa)G_{s,L}^2(\kappa, z) - 2N_n(\gamma_z\kappa)N_n(\alpha_z\kappa)G_{s,L}^2(\kappa, z)J_0(\kappa z\theta_L)]$$

(76)

Those above the LGS beacon are same as Eq. (17) and Eq. (24).

In above equations, γ_z, θ_N, and $G_{s,N}$ (or α_z, θ_L, $G_{s,L}$) are main related parameters of anisoplanatic effect, and they are the propagating factor, the offset angle, and the filter function of the NGS (or LGS), respectively. Here we have supposed that the altitude of NGS is same as that of the target.

Using these filter functions, the effective anisoplantic variance for this particular LGS AO system can be calculated and expressed as follows:

$$\sigma_{\text{eff}}^2 = (\sigma_{1,1;N}^2 + \sigma_{2,0;N}^2) + [\sigma_{\phi,L}^2 - (\sigma_{0,0;L}^2 + \sigma_{1,1;L}^2 + \sigma_{2,0;L}^2)]$$

(77)

In this equation, the first two items in parentheses are the contribution of the NGS subsystem, describing the anisoplanatism of tip-tilt and defocus modes respectively. While the items in brackets are the contribution of the LGS subsystem, and the four items are the variance of the total phase, the piston, the tip-tilt and the defocus mode, sequentially. As a comparison, the effective anisoplanatic variance for a usual LGS AO system, in which only tip-tilt mode can be sensed and corrected by the NGS subsystem, can be expressed as:

$$\sigma_{\text{eff}}^2 = \sigma_{1,1;N}^2 + \left[\sigma_{\phi,L}^2 - \left(\sigma_{0,0;L}^2 + \sigma_{1,1;L}^2\right)\right]$$

(78)

Obviously, for this special LGS AO system, the contribution of the defocus mode to the effective anisoplanatic variance comes from the NGS system, i.e $\sigma_{2,0;N}^2$, while for a usual LGS AO system, it comes from the LGS subsystem.

Below we give some numerical results. We mainly study the changes of the anisoplanatic variance with some control parameters, including the altitudes (L and H), the offset angles (θN and θL), and the angular width (for Gaussian sources: $\theta r,N$ and $\theta r,L$) of the NGS and LGS sources. Some typical results are showed in the figures below. In our calculation, the altitude of the target L is selected as 500km500km (a typical value for a LEO satellite), and the wavelength as 1.315μm1.315μm.

In Figure 7(a) and Figure 7(b), the changes of the anisoplanatic variance with the angular widths and the offset angles of the beacons are given. In this case the invalid piston component of variance has been removed. In these figures, we also compare the values for three different altitudes of beacons, including a NGS (H=L=500km) and two kinds of LGSs with altitude H=15km and H=90kmrespectively. It is easy to see that the variances generally increase with the offset angles and the angular widths of the beacons. But there is some minor difference for the beacon size: the variance first decrease as beacon size increases, then it increases. We can also see that the changes are more obvious when the altitudes of the beacons are larger, for example, we can see the variance changes from 0.1 to 1.6 rad2rad2 when the offset angles changes from 0 to 10››10›› for NGS, but there are nearly no changes for 15km Rayleigh LGS, as showed in Figure 7(b).

In Figure 7(c) and (d), the components of anisoplanatic variance below and above the beacon, are given respectively. The values for the total phase and its first three components (piston, tilt and defocus) are showed altogether. In Figure 7(e), the variances for the total phase, the piston and tip-tilt removed phase, and the piston and tip-tilt and defocus removed phase, are showed respectively. When the altitudes of the beacon are more than 20 km the variances are almost the same as the results of the NGS. In Figure 7(f), the effective anisoplanatic variances expressed by Eq. (77) are showed for three different offset angles of NGS.

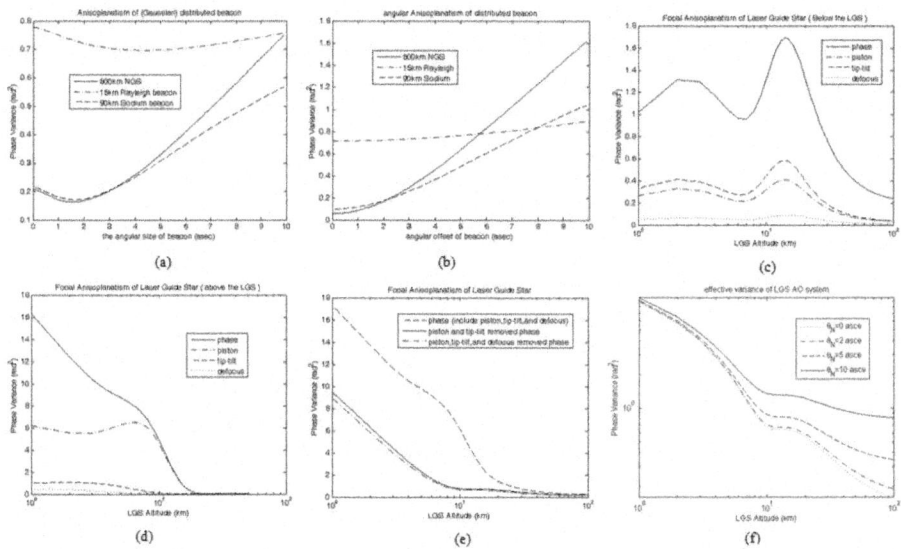

Figure 7: (a) Anisoplanatism of distributed beacon; (b) angular anisoplanatism; (c) the focal anisoplanatism (below the beacon); (d) focal anisoplanatism (above the beacon); (e) focal anisoplanatism (sum); (f) effective variance.

For the special LGS AO system, the anisoplanatic variance of defocus comes from NGS sub-system and not from the LGS subsystems as usual LGS AO systems. In Figure 8, we compare the values of these two variances and the relationship between the altitude of LGS and the offset angle of NGS. The transverse coordinates are magnitudes of the variances. The solid line describes the change of the defocus variances with the altitude of LGS and the altitude of LGS is showed in the left longitudinal coordinates. Similarly, the dotted line describes the change of the defocus variances with the offset angle of NGS and the offset angle of NGS is showed in the right longitudinal coordinates.

From this figure the value of the LGS altitude and the NGS offset angle, having the same value of the variance, can be read directly and some operational conclusions can be drawn. For example, for a Rayleigh LGS (with an altitude of 10km to 20km) the anisoplanatic variance of the focus component has the value between 0.08 to 0.1 rad 2, same as that for a NGS with the offset angle between S›8›› and››9››. Similarly, the sodium LGS (with altitude of 90km) correspond to the offset angle of NGS between››2›› and ‹›3››. It is also easy to see that the variance is a monotonically increasing function of the NGS offset angle and a almost monotonically decreasing function of the LGS altitude. Therefore, if the NGS offset angle is smaller or ‹›0›› (such as directly imaging of a bright satellite) using NGS to correct the defocus component, the variance is smaller. Otherwise, when the NGS offset angle is larger (for example, when projecting laser beams to a LEO satellite, the advance angle about ‹›10›› must be considered) using sodium LGS to correct the defocus the variance is smaller.

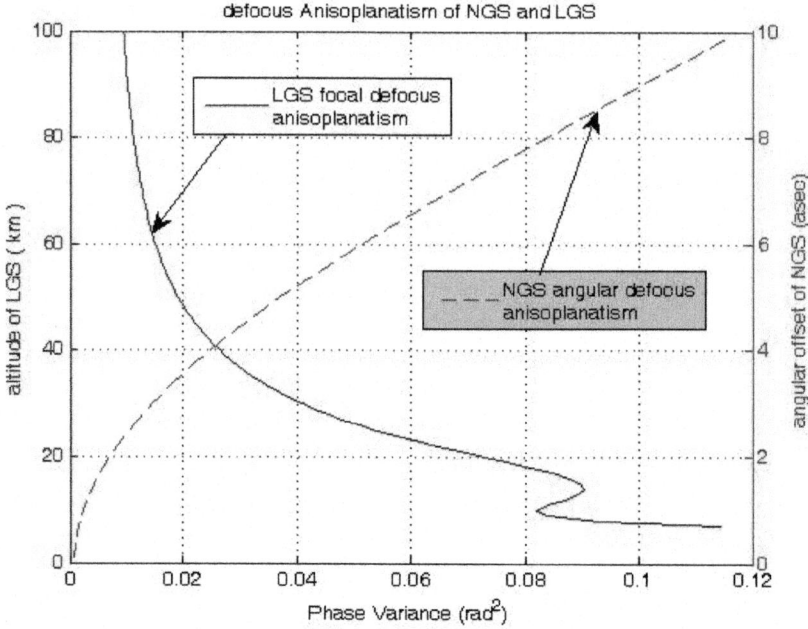

Figure 8: The anisoplanatism of the defocus component

SUMMARY

Using transverse spectral filtering techniques we reconsider the anisoplanatism of general AO systems. A general but simple formula was given to find the anisoplanatic variance of the turbulence-induced phase and its arbitrary Zernike components under the general geometry of AO systems. This general geometry can describe most kinds of anisoplanatism appearing in currently running AO systems, including angular anisoplanatism, focal anisoplanatism and that induced by distributed sources or separated apertures, and so on. Under some special geometry, close-form solutions can be obtained and are consistent with classic results, which prove the effectiveness and universality of the general formula constructed in this chapter. We also give some numerical results of hybrid anisoplanatism under some more complex geometry.

REFERENCES

1. L. C Andrews, 1998Special functions of mathematics for engineers. SPIE-International Society for Optical Engineering, Bellingham

2. D. F Buscher, G. D Love, et al2002Laser beacon wave-front sensing without focal anisoplanatism. Opt. Lett., 273149151

3. J. O Dickey, P. L Bender, et al1994Lunar laser ranging- A continuing legacy of the Apollo program. Science 2655171482490

4. S Esposito, R Ragazzoni, et al2000Absolute tilt from a laser guide star: a first experiment. Experimental Astronomy 101135145

5. S Esposito, A Riccardi, et al1996Focus anisoplanatism effects on tip-tilt compensation for adaptive optics with use of a sodium laser beacon as a tracking reference. J.Opt.Soc.Am.A, 13919161923

6. R Foy, A Migus, et al1995The polychromatic artificial sodium star: A new concept for correcting the atmospheric tilt. Astron. Astrophys. Suppl. Ser., 1113569578

7. D. L Fried, 1982Anisoplanatism in adaptive optics. J.Opt.Soc.Am, 7215261

8. D. L Fried, 1990Time-delay-induced mean-square error in adaptive optics. J.Opt.Soc.Am.A 7712241225

9. D. L Fried, 1995Focus anisoplanatism in the limit of infinitely many artificial-guide-star reference spots. J.Opt.Soc.Am.A, 125939949

10. D. T Gavel, J. R Morris, et al1994Systematic design and analysis of laser-guide-star adaptive-optics systems for large telescopes. J.Opt.Soc. Am.A,112914924

11. W Happer, G. J Macdonald, et al1994Atmospheric-turbulence compensation by resonant optical backscattering from the sodium layer in the upper atmosphere. J.Opt.Soc.Am.A, 111263276

12. J. W Hardy, 1998Adaptive optics for astronomical telescopes. Oxford University Press.

13. F Marc, H. G De Chatellus, et al2009Effects of laser beam propagation and saturation on the spatial shape of sodium laser guide stars. Optics Express 17749204931

14. N Muller, V Michau, et al2011Differential focal anisoplanatism in laser guide star wavefront sensing on extremely large telescopes. Opt. Lett., 362040714073

15. C. R Neyman, 1996Focus anisoplanatism: A limit to the determination of tip-tilt with laser guide stars. Opt. Lett., 212218061808

16. R. R Parenti, and R. J Sasiela, 1994Laser-guide-star systems for astronomical applications. J.Opt.Soc.Am.A, 111288309

17. R Ragazzoni, B Le, Roux, et al. (2005Multi-Conjugate Adaptive Optics for ELTs: constraints and limitations. C. R. Phys. 61010811088

18. S Riepl, W Schluter, et al1999Evaluation of an SLR adaptive optics system. Laser Radar Ranging and Atmospheric Lidar Techniques II. U. Schreiber and C. Werner. Bellingham. Proc. SPIE. 38659095

19. F Rigaut, and E Gendron, 1992Laser guide star in adaptive optics-The tilt determination problem. Astron.Astrophys. 2612677684

20. R. J Sasiela, 1992Strehl ratios with various types of anisoplanatism. J.Opt.Soc.Am.A, 9813981405

21. R. J Sasiela, 1994Wave-front correction by one or more synthetic beacons. J.Opt.Soc.Am.A, 111379393

22. R. J Sasiela, 2007Electromagnetic wave propagation in turbulence: evaluation and application of Mellin transforms. SPIE Press.

23. R. J Sasiela, and J. D Shelton, 1993Transverse spectral filtering and Mellin transform techniques applied to the effect of outer scale on tilt and tilt anisoplanatism. J.Opt.Soc.Am.A, 104646660

24. P. D Stroud, 1996Anisoplanatism in adaptive optics compensation of a focused beam with use of distributed beacons. J.Opt.Soc.Am.A, 134868874

25. G. A Tyler, 1994Bandwidth considerations for tracking through turbulence. J.Opt.Soc.Am.A, 111358367

26. Tyler, G. A. (1994). Rapid evaluation of d 0 - the effective diameter of a laser-guide-star adaptive-optics system. J.Opt.Soc.Am.A, 11(1): 325-338.

27. G. A Tyler, 1994Wave-front compensation for imaging with off-axis guide stars. J.Opt.Soc.Am.A, 111339346

28. R Tyson, 2011Principles of adaptive optics (3rd Edition ed.), CRC Press, Boca Raton.

29. E. P Wallner, 1977Minimizing atmospheric dispersion effects in compensated imaging. J.Opt.Soc.Am., 673407409

30. B. M Welsh, and C. S Gardner, 1991Effects of turbulence-induced anisoplanatism on the imaging performance of adaptive-astronomical telescopes using laser guide stars. J.Opt.Soc.Am.A, 816980

31. M. R Whiteley, M. C Roggemann, et al1998Temporal properties of the Zernike expansion coefficients of turbulence-induced phase aberrations for aperture and source motion. J.Opt.Soc.Am.A, 1549931005

32. M. R Whiteley, B. M Welsh, et al1998optimal modal wave-front compensation for anisoplanatism in adaptive optics. J.Opt.Soc. Am.A,15820972106

33. K. E Wilson, 1994An overview of the Compensated earth-Moon-earth laser link (CEMERLL) experiment. Bellingham. Proc. SPIE. 21236674

Chapter 11

REINVENTING QUANTUM PHYSICS

Jean-Paul Auffray[1], Mohamed S. El Naschie[2]

[1]Ex Courant Institute of Mathematical Sciences, New York University, New York, NY, USA

[2]Department of Physics, University of Alexandria, Alexandria, Egypt

ABSTRACT

Quantum Physics (QP) was invented in the early years of the Twentieth century by physicists born and educated in the western world. We examine the possibility that this is the main reason—or at least one of the main reasons—which caused QP to go astray from the start. We present the ABC for a renovated Quantum Physics.

INTRODUCTION

One question currently haunts the minds of physicists around the world whatever their fields of expertise might be—aerospace science, computational mathematics, operations research, astronomy, astrophysics, quantum physics.... It arises as follows.

As human beings we experience the irresistible certainty of living (existing) in a world composed of four constitutive ingredients: space, time, matter and energy. To us, furthermore, space has three dimensions—width, height and depth—, time flows evenly in one dimension—yesterday, today, tomorrow—. Matter is made up of molecules, themselves made up of atoms, themselves made up of "elementary particles", themselves made up of...

Of what?

Of points. Points, whatever they are, seem to be the ultimate fabrics of the universe—the cosmos, the System of the World, call it as you wish—our world.

What about energy?

It seems to exist in the universe in two distinct forms: "ordinary energy" and "dark energy".

Dark energy? What is it, where is it, and why is it "dark"…?

We examine these three disturbing questions in this note.

THE PROBLEM WITH DARK ENERGY

Behind the appearances noted above, lies of course the mysterious world of the "quantum".

Quantum Physics (QP) was invented in the early years of the twentieth century by physicists born and educated in the western world. They were taught to write the equations which occur in their calculations so as to be read from left to right, as for example this (modest) equation

$$1 \times 1 = 1. \tag{1a}$$

Ancient Egyptians had more imagination. Their hieroglyphic inscriptions can be read from left to right, but also, for some of them, from right to left, or from top to bottom or bottom to top (Figure 1).

To proceed further along this line of reasoning, we placed ourselves under the protection (the guidance) of the Goddess Maat said to have regulated the stars, the seasons, and the actions of mortals and deities in Ancient Egypt, and to have set the order of the universe from chaos at the moment of creation— precisely a question we investigated in preceding notes.

By Egyptian conventions, when Maat looks to her right as in Figure 1(a), she expects to be approached from the left. We approached her from the left and we presented her with our Equation (1a) as a playful riddle.

She inspected the riddle and told told us with an inquisitive smile on her face (our interpretation, our translation): "Written as you have written it, your riddle expresses a modest arithmetic "truth" of the kind that a well-trained elementary western world school-teacher might want to transmit to the young children he/she has the charge of educating

We returned to Maat the next day. To our surprise, she was looking to her left this time, (to the right for us as on Figure 1(b)). She told us, again with a friendly greeting smile: "Do you remember the riddle you presented to me yesterday?" We said, yes we remembered it. She said: "Now read the riddle from right to left." We did. And we felt frustrated. What we saw—when converted visually to our usual way of reading in the western world, i.e. left to right—was:

$$1 = 1 \times 1,$$

(1b)

Maat said: "Do not feel frustrated. Written this way the riddle now says that the entity "1" is a "composite"—here is the "product"—of two "units", a statement more "philosophical" than the one you (we) derived from the riddle yesterday."

We were startled!

And even more so when Maat added with a serious look on her face: "And seize this opportunity to put an end to your use, abuse and misuse in your writings of the word quantum."

Figure 1: How to read hieroglyphic inscriptions.

http://emmahardieacientcivilisations.weebly.com/uploads/3/0/3/6/30367743/4814950.jpeg?1403579135.

Figure 1(a): .Maat looks to her right when she expects to be approached from the left (her right).

https://sp.yimg.com/ib/th?id=JN.WMQRtes99bTyXuuZQDXrMw&pid=15.1&P=0.

Figure 1(b): Maat looks to her left when she expects to be approached from the right (her left).

https://sp.yimg.com/ib/th?id=JN.WMQRtes99bTyXuuZQDXrMw&pid=15.1&P=0.

We received this "order" from Maat as an injunction. Keeping in mind that an injunction is "an equitable remedy in the form of a court order that compels a party to do or refrain from specific acts" we decided to comply with it at once without further ado.

A HUGE PROBLEM

We collected the information available concerning the current use of the word quantum in theoretical plysics. We came up with this short list.

1)   Energy in the the universe is not "quantized". Motion is.

2)   Motion occurs in nature in the form of elements each carrying the same amount, the same quantity—the same quantum—of motion. NB In this statement we substituted the word motion, easily understood, for the word action formally used in this context but largely ignored by physicists at large, one of them Albert Einstein, who throughout his life throrougly ignored its use and introduced instead the faulty concept of "energy quanta".

3)   As suggested by French mining engineer genius mathematician Henri Poincaré shortly before his premature death in 1912, the motion element—the quantum—contains points "which are equivalent to one another from the standpoint of probability", said Poincaré [1] .

This raises a huge problem: Poincaré tells us that the motion element—the quantum—contains points. Now if the quantum contains points, whatever they might be, how does it maintain its integrity? This would be like trying to keep together as a unit a group of tourists visiting some shrine abroad!

We had to find a way out.

We searched… and came up with this solution.

TOWARD A SOLUTION

To describe the hidden functioning of the quantum in the System of the World, we developed a theoretical scheme in the framework of which, when written western style, two of the fundamental equations read [2] :

$$lp = Ed = h. \tag{2a}$$

Never mind at the moment what the symbols these equations contain represent. If, instead of being read left to right these equations are read right to left (but printed left to right so as to be read "normally" by western physicists), they would read

$$h = dE = pl. \tag{2b}$$

Let us explore their meaning when written this way.

WHAT TO DO WHEN YOU MEET AN UNKNOWN

Arab scholars imported to the territories they had conquered on the Spanish Peninsula during the Middle Age their practice of algebra ("al-jabr", to them) meaning "reunion of broken parts". It allowed them to convert typically an equation such as $3 - 1 = 2$ into $3 = 2 + 1$, thereby reuniting the "broken parts". To represent an unknown entity in this art, they used, not a graphic, but a "sound", which became the sound of the Greek letter χ and finally the Latin letter x that we still use today to designate the unknown in an algebraic equation.

Let us call "x" instead of "h" the unknown value that the composites lp and Ed have in common according to Equations (2b). These equations now read

$$x = lp \tag{2c}$$

$$x = Ed$$

For the sake of curiosity, which sometimes is a key to discovery, let us use an Egyptian hieroglyph to represent this common value, for example the hieroglyph:

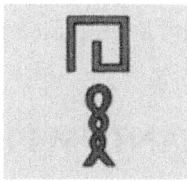

which suggests some form of question mark, some form of ambiguity, of perplexity—of uncertainty, of inquisitiveness. And while we are at it... let us write Equations (2c) using hieroglyphs throughout, an easy task since a well-defined hieroglyph exists for each of the four letters, l, p, E and d the equations contain. We show the result in Figure 2(a) and Figure 2(b).

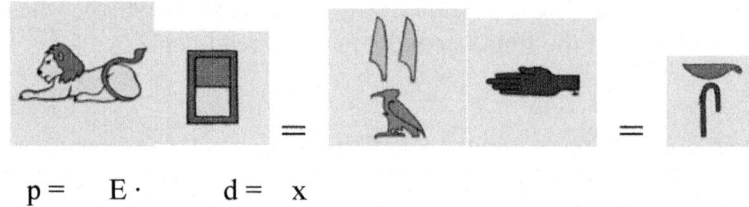

l ·   p =   E ·   d =   x

Figure 2(a): (a). lp = Ed = x written with hieroglyphs read left to right.

http://www.artyfactory.com/egyptian_art/egyptian_hieroglyphs/images/icons/l.gif.

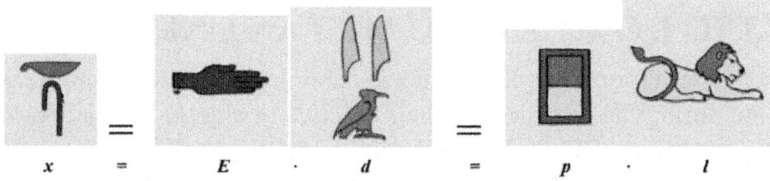

x   =   E   ·   d   =   p   ·   l

Figure 2(b): (b). lp = Ed = x written with hieroglyphs read right to left.

http://www.artyfactory.com/egyptian_art/egyptian_hieroglyphs/images/icons/l.gif.

We remembered one significant word of the English language not often used. A statement is said to be informative if it expresses a proposition containing just one alternative possibility. General Relativity, for instance, is an informative theory: it asserts that gravitation results from curvatures induced by matter and energy in a 4- dimensional spacetime. It asserts something… which is not necessarily "true". By contrast, when properly understood, Equations (2c) constitute an inquisitive statement—they express a proposition containing two or more alternative possibilities.

All this having been said and attempted let us affront without further delay the moment of truth.

REFORMULATING QUANTUM PHYSICS

Discard the Symbol h Representing the Planck Constant

The systematic misuse and abuse by physicists of the symbol h ever since its invention by Max Planck in 1900 to represent the elementary quantum of action in equations have made of this "constant" a handicap for the sane development of Quantum Physics. Let us respectfully place it in the Quantum Museum of Obsolete Concepts, the QMOC.

Incidentally, as we have shown in a preceding note [2] , Albert Einstein himself made a valiant effort to get rid of the Planck constant in the first major paper he wrote during his Annus mirabilis (1905) [3] . This is not a loss anyway since this "constant" is not a constant in the strict sense of the word and cannot legitimately enter as a factor in algebraic equations.

Acknowledge the Quantum as Being a Ubiquitous Active Principle

Undertstood to constitute a (mathematical) statement describing a physical reality, Equations (2c) tell us that the (physical) entity represented by the symbol x can express itself, or be expressed, in the form of at least two composites, lp and Ed. This confers to this entitity the character of inquisitivness: one question, at least two different answers. Inquisitive: they can be read from left to right, we call the x as representing the Xon, or from right to left, we call it as representing the noX. We now claim that inquisitive, the quantum is also ubiquitous (from the Latin ibique, meaning "everywhere"), i.e. it is or seems to be everywhere at the same time—it is omnipresent in French,ubicuo in Spanish, alomtegenwoordig in Dutch, 无处不在in Chinese, вездесущий in Russian (Figure 3).

As it "expresses itself"—as it "occurs"—in the System of the World (the Void; nature, the cosmos) the Xon/noX quantum generates (liberates) continuouly, but discontinuously, points—shall we call them quanta?

Returning to our starting point in this note, we are now ready to place the cherry on the cake (Figure 4).

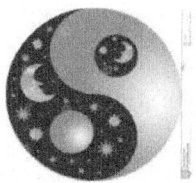

Figure 3: An inquisitive ambiguous ubiquitous active principle runs the System of the World.

https://encrypted-tbn2.gstatic.com/images?q=tbn:ANd9GcQ-vcqN2rvQ-ameh9LjazpBy8jmHMB jDL3NqxrHTeFYwn6fny8_kw

Figure 4: The cherry on the cake. The fallacious Planck constant is gone.

http://www.devotedtocakes.ie/wp-content/uploads/2014/02/Chinese-New-Year-Cake.jpg

Place the Cherry on the Cake

By the scheme exposed in this note, the quantum expresses itself in the System of the World by generating in the Void composites of the form ab, with a and b forming, as of this writing, four known possible combinations [2] :

$$x = lp, \ Ed, \ \varphi\sigma \text{ and / or } e\chi$$

where

l measures a fractal space length,

p a linear momentum,

E an energy lasting for a time duration d,

φ measures an angle,

σ the conjugate angular momentum,

e an (electromagnetic) charge and

χ the corresponding (mysterious) gauge function.

This being, we see that the quantum—the Xon/noX—generates in the Void the ingredients constitutive of space, momentum, energy, time and more.

CONCLUSIONS

Interestingly, what we reported in this note coincides with the contents and the spirit of statements formulated recently by one of us [4] —using a more complex vocabulary—stating for example: "The preceding explanation amounts to a paradigm shift in physics where the totally empty vacuum of spacetime is taken as fundamental and everything else is derivable from it", or "All forms of energy and matter represented by [the] iconic equation E = mc² are nothing but the zero point energy fluctuations of the real vacuum of spacetime" (our emphasis).

And so, where is dark energy located in the universe?

As per the reasonings developed in the present note, energy—be it ordinary or dark—is generated randomly and discontinuously in the Void by the quantum. And thus, in this sense, it is fair to say that "the Phantom (energy, ordinary and dark) is in the House"—and it is not constituted of "energy quanta".

ACKNOWLEDGEMENTS

We are indebted to Ms. Clara Gao, to Ms. Kelly Sang and to Ms. Freya Zhang, JMP Editorial Board Assistants, for their valuable help and kind advice in properly formatting this paper for publication in this journal.

REFERENCES

- Poincaré, H. (1963) Mathematics and Science: Last Essays. Dover, New York.
- Auffray, J.-P. (2015) Journal of Modern Physics, 6, 1478-1491. http://dx.doi.org/10.4236/jmp.2015.611152
- Einstein, A. (1905) Annals of Physics, 17, 132. http://www.esfm2005.ipn.mx/ESFM_Images/paper1.pdf
- El Naschie, M.S. (2015) Open Journal of Applied Sciences, 5, 313-324. http://dx.doi.org/10.4236/ojapps.2015.57032

CITATION

CHAPTER 1

Martin Veis and Roman Antos (2012). Atomic Force Microscopy in Optical Imaging and Characterization, Atomic Force Microscopy - Imaging, Measuring and Manipulating Surfaces at the Atomic Scale, Dr. Victor Bellitto (Ed.), ISBN: 978-953-51-0414-8

CHAPTER 2

Duc Doan Hong and Fushinobu Kazuyoshi (2012). Fluidic Optical Devices Based on Thermal Lens Effect, Optical Devices in Communication and Computation, Dr. Peng Xi (Ed.), ISBN: 978-953-51-0763-7, InTech, DOI: 10.5772/48072.

CHAPTER 3

Kyung M. Choi (2012). Novel Optical Device Materials - Molecular-Level Hybridization, Optical Devices in Communication and Computation, Dr. Peng Xi (Ed.), ISBN: 978-953-51-0763-7, InTech, DOI: 10.5772/50032.

CHAPTER 4

Luxi Li, Xianbo Shi, Cherice M. Evans and Gary L. Findley (2012). Atomic and Molecular Low-n Rydberg States in Near Critical Point Fluids, Advanced Aspects of Spectroscopy, Dr. Muhammad Akhyar Farrukh (Ed.), ISBN: 978-953-51-0715-6, InTech, DOI: 10.5772/48089.

CHAPTER 5

Christensen Jr., W. (2015) Relativized Quantum Physics Generating N-Valued Coulomb Force and Atomic Hydrogen Energy Spectrum. *Journal of Modern Physics*, **6**, 194-200. doi: 10.4236/jmp.2015.63025.

CHAPTER 6

Z. Kong, "The Atomic Regular Polyhedron Electronic Shell," *Journal of Modern Physics*, Vol. 4 No. 10A, 2013, pp. 1-19. doi: 10.4236/jmp.2013.410A1001.

CHAPTER 7

Dorado, M. (2014) Molecular Beam Depletion: A New Approach. *Journal of Modern Physics*, **5**, 1139-1145. doi:10.4236/jmp.2014.512116.

CHAPTER 8

Zoran Popovic, Jörgen Thaung, Per Knutsson and Mette Owner-Petersen (2012). Dual Conjugate Adaptive Optics Prototype for Wide Field High Resolution Retinal Imaging, Adaptive Optics Progress, Dr. Robert Tyson (Ed.), ISBN: 978-953-51-0894-8, InTech, DOI: 10.5772/53640.

CHAPTER 9

S. Bonora, R.J. Zawadzki, G. Naletto, U. Bortolozzo and S. Residori (2012). Devices and Techniques for Sensorless Adaptive Optics, Adaptive Optics Progress, Dr. Robert Tyson (Ed.), ISBN: 978-953-51-0894-8, InTech, DOI: 10.5772/53550

CHAPTER 10

Jingyuan Chen and Xiang Chang (2012). A Unified Approach to Analysing the Anisoplanatism of Adaptive Optical Systems, Adaptive Optics Progress, Dr. Robert Tyson (Ed.), ISBN: 978-953-51-0894-8, InTech, DOI: 10.5772/54602.

CHAPTER 11

Jean-Paul Auffray, Mohamed S. ElNaschie, (2016) Reinventing Quantum Physics. *Journal of Modern Physics*, 07, 156-161. doi: 10.4236/jmp.2016.7101

INDEX